Majan International Conference (MIC 2018)

Promoting Entrepreneurship and Technological Skills: National Needs, Global Trends

Muscat, Oman
19 March 2018

Editors:

Dr. Brian Poole **Dr D. Ramalingam**

ISBN: 978-1-5108-6299-9

Printed from e-media with permission by:

Curran Associates, Inc.
57 Morehouse Lane
Red Hook, NY 12571

Some format issues inherent in the e-media version may also appear in this print version.

Printed by Curran Associates, Inc. (2018)

For permission requests, please contact Majan University College
at the address below.

Majan University College
Mutrah | Way No. 2621 | Building No. 1986
P. O. Box: 710, P.C.: 112
Ruwi | Sultanate of Oman

Phone: +968 2473 0400

info@majancollege.edu.om

Additional copies of this publication are available from:

Curran Associates, Inc.
57 Morehouse Lane
Red Hook, NY 12571 USA
Phone: 845-758-0400
Fax: 845-758-2633
Email: curran@proceedings.com
Web: www.proceedings.com

TABLE OF CONTENTS

Preface

Majan International Conference 2018 (MIC 2018) was organised by Majan College (University College) (MCUC), Muscat, Sultanate of Oman during 19–20 March 2018 at its campus. This is the second International conference organised by MCUC.

MCUC is one of the prime higher education institute in Muscat, Sultanate of Oman established in the year 1995 and it's also the first private University College in the country. MCUC is providing quality higher education in Business, English and Information Technology fields with academic affiliation with University of Bedfordshire, UK.

MIC 2018 had the broader theme **"Promoting Entrepreneurship and Technological Skills: National Needs, Global Trends**". The conference provided a forum for national, regional and international experts, academics, policy-makers and thought leaders to express and discuss their views on the role of entrepreneurial and technological skills in diversifying and expanding Oman's economy. The subthemes of the conference are ground under three categories such as Theme 1 :Entrepreneurship, Business and Management, Theme 2:Information Technology and Theme 3: Soft Skills, Enabling Skills with the following titles: Generating business ideas; entrepreneurial business planning; entrepreneurship and economic growth; entrepreneurial case studies; corporate entrepreneurship/commerce and e-retail; fostering creativity and innovation; Technopreneurship; technology-based enterprises; intelligent informatics; smart cities and smart societies; computational thinking; new media literacy; soft skills and entrepreneurship; 21st century communication skills; Intercultural competence and global awareness; and problem-solving skills.

MIC 2018 received original unpublished articles from 352 authors spread across 21 Nations across the globe. MIC 2018 organizing committee formed a programme committee with 116 international subject experts possessing a minimum PhD qualification and strong research profile. All the full papers and abstract only submissions were screened for plagiarism and double-blind reviewed using EasyChair conference management system. Each paper was reviewed by at least two and at most five reviewers. An average of 38 % papers were selected for presentation under all three themes.

The four internationally renowned key note speakers shared their experience and views on topics such as "***Entrepreneurship Ventures", "promoting a culture of high performance, entrepreneurship and innovation"***, ***"data security and business intelligence",*** and "***The Key to Active Online Learning***" to the audience.

Selected papers presented at the conference were accepted for publication by the publication partners IEEE Xplore **(ISBN: 978-1-5386-3761-6)** and other accepted papers are published in this proceedings.

Dr Maha Kobeil
Conference General Chair,
Majan International Conference 2018.

Dr Geetha Achuthan ,Caledonian College of Engineering ,Oman
Dr Girija Narasimhan ,Higher College of Technology ,Oman
Dr Gn Balaji GN ,Cvr College of Engineering, Hyderabad ,India
Dr Gregory Epiphaniou ,University of Wolverhampton ,UK
Dr Hamda Darwish Al-Shihi ,Omani Society For Educational Technology. ,Oman
Dr Huda Al Shuaily ,Higher College of Technology ,Oman
Dr Hussein Fakhry ,Zayed University ,UAE
Dr Iain Mcgee ,Majan College (University College) ,Oman
Dr Intisar Al-Mejibli ,University of Information Technology And Communications ,Iraq
Dr Jabar H. Youif ,Sohar University ,Oman
Dr Jane Eastham ,Harper Adams University ,UK
Dr Javeed Hussain ,Global College of Engineering And Technology ,Oman
Dr Jeevanandam Jeevanandam J ,Hkbk College of Engineering ,India
Dr K.Malar K.Malar Mathi ,Bharathiar University ,India
Dr Kabaly P Subramanian ,Arab Open University ,Oman
Dr Kallam Suresh , China University of Technology,Nanchang ,China
Dr Khamis Siad Al Muniri ,University of Buraimi ,Oman
Dr Labeeb Zeeshan ,College of Banking And Financial Studies ,Oman
Dr Madhuri Goswami ,Government Pg Girls College ,India
Dr Mahdi Bahaghighat ,Amirkabir University of Technology ,Iran
Dr Mallika Elancheran ,Salalah College of Technology ,Oman
Dr Mansour Naser Alraja ,Dhofar University ,Oman
Dr Mohammad Mohatram ,Waljat College of Applied Sciences ,Oman
Dr Mohammad Alazawi ,Oman College of Management And Technology ,Oman
Dr Mohammed Aref Abdul Rasheed ,Dhofar University ,Oman
Dr Muhammad Saqib ,Middle East College ,Oman
Dr Nirmal Gunupudi ,Majan College (University College) ,Oman
Dr Osman Ahmed El-Said ,German University of Technology In Oman ,Oman
Dr Padmapriya A , Alagappa University ,India
Dr Poonam Kumar ,Pearl Academy ,India
Dr Ramalingam Dharmalingam ,Majan College (University College) ,Oman
Dr Ramanujam Velusamy ,Bsmed, Bharathiar University ,India
Dr Rizwan Patan ,Vit University ,India
Dr Robina Ahmed ,Modern College of Business Studies ,Oman
Dr Rolando Lontok Jr ,Nizwa College of Technology ,Oman
Dr Rolou Lyn Maata ,Gulf College Muscat Sultanate of Oman ,Oman
Dr Rowena Ocier Sibayan Ed. ,Gulf College ,Oman
Dr Rula Al Azawi ,De-Montfort University ,UK
Dr Salah Al Majeed ,Ieee Oman Section ,Oman
Dr Saleh Babaa ,Military Technological College ,Oman
Dr Salman Nusrat ,College of Banking And Financial Studies ,Oman
Dr Salman Saadat ,Military Technical College ,Oman
Dr Samee Sayyad ,Marathwada Institute of Technology (Mit) ,India
Dr Seema Arora ,Waljat College of Applied Sciences ,Oman
Dr Shadha Alamri ,Higher College of Technology ,Oman
Dr Shaik Asif Hussain ,Middle East College ,Oman
Dr Shantanu Shekhar ,Al Musanna College of Technology ,Oman
Dr Smitha Sunil Kumaran Nair ,Middle East College ,Oman
Dr Sohail Iqbal Malik ,Buraimi University College ,Oman
Dr Suhail Mohammad Ghouse ,Dhofar University ,Oman
Dr Sv Suryanarayana ,Cvr College of Engineering ,India
Dr Tahseen Arshi ,Majan College (University College) ,Oman
Dr Tarek Tarek, Gulf College ,Oman
Dr Tariq Husain, Military Technical College ,Oman
Dr Tauqeer Safdar ,Universiti Teknologi Petronas ,Malaysia
Dr Tripti Sharma, Higher College of Technology,Oman

Dr Udhayanan Pappusamy ,Majan College (University College) ,Oman
Dr Umaselvi Ashogan ,Majan College (University College) ,Oman
Dr Vaishnavi D ,Vardhaman College of Engineering ,India
Dr Veena Tewari ,Majan College (University College) ,Oman
Dr Vidhya Lavanya ,Middle East College ,Oman
Dr Vijaya Padmanabha ,Waljat College of Applied Sciences ,Oman
Dr Vinay Gupta, Muscat College ,Oman
Dr Viswanathan Thangaraj ,Symbiosis Institute of Business Management ,India
Dr Vladimir Dyo ,University of Bedfordshire ,UK
Dr Wasi Uz Zaman. Khan ,Military Technological College ,Oman
Dr Wilfred Blessing N.R ,Annai Vailankanni College of Engineering ,India
Dr Zied Othman ,Mustansiriyah University ,Iraq
Prof. Mark Mabey ,New Beacon Group ,UK
Prof. Tony Ward ,University of York ,UK
Prof. Venkoba Rao ,Majan College (University College) ,Oman
Prof. Vijay Nair ,Majan College (University College) ,Oman
Prof. Ahmed T. Sadiq ,University of Technology ,Iraq
Prof. Andrew Ordys , Ieee Oman Section, Military Technological College ,Oman
Prof. Dileep Kumar Mohanachandran, Berjaya University College, Kuala Lumpur, Malaysia
Prof. Marikkannan M ,Institute of Road And Transport Technology, ,India
Prof. Munawar Hameed, International Applied Education Research Centre, Pakistan
Prof. Paul Burns, University of Bedfordshire, UK

DOES EMPLOYEE CREATIVITY COME THROUGH TRANSFORMATIONAL LEADERSHIP AND ORGANIZATIONAL SUPPORT?

Muhammad Iftikhar Ali
Bahria University, Islamabad Pakistan
iftikhar.kfp@gmail.com

ABSTRACT

The role of transformational leadership and organizational support in achieving employee's creativity has gained much attention in research and practice. However, previous studies did not indicate any influence of transformational leadership over employees' creativity when they have organizational support in Pakistan telecommunication sector. This study considered organizational support and four dimensions of transformational leadership include idealized influence, inspirational motivation, intellectual stimulation, individual consideration and it was found that the combined effect of these dimensions influences employee's creativity. The sample size of the study was 260. In this quantitative research survey was designed to gather data from respondents. Transformational leadership theory and organizational support theory is well supported by this study. SPSS was used to analyze the data. The results gained by conducting this study confirm the findings of earlier studies which illustrate the significant relationship among transformational leadership, organizational support and employee's creativity. Findings of this study provide avers for leadership practice, especially both for leaders and their followers to interact with each other.

Keywords: Transformational Leadership, Organizational Support, Employees Creativity.

INTRODUCTION

An organization is a group of people work and execute together in order to reach common goals and objectives through individual and group efforts. The organizational objectives are achieved through organizational support (Mehraban & Avanloo, 2014). However, employee creativity is also a factor that may affect the output of employees in telecom sector (DeConinck & Johnson, 2009). No doubt, transformational leadership is also much important for organizational success. It has been focused more on vision oriented process and attained the major level of research in last few years. It is the sole duty of leader to craft an environment in which his followers work in same manners towards the attainment of specific goals (Jyoti & Dev, 2015). It is the responsibility of transformational leaders to drag the interests of employees from personal to collective goals (Mittal & Dhar, 2015) and provoke them to put extra beyond their duties (Jyoti & Dev, 2015). Transformational leadership manages the innovation and outcome of their subordinates and persist other to research on different dimensions of novelty, loyalty and outcome (Mittal & Dhar, 2015). It has four divisions: approaching first to idealized influence/ charisma, leaders toil to develop sense of combined mission, homage, loyalty and motivation (Shin & Zhou, 2003). In individualized consideration leaders toil to craft closely associated relation with their subordinates (Jung, Chow & Wu, 2003) and tried to solve their issues alongside evaluating their skills and ambitions in addition to mentoring them in way that they are acquainted with confidence in themselves with picture of imminent challenges (Shin & Zhou, 2003). Intellectual stimulation is somehow more pertinent to innovation involved in problem solving as well as ascending and broadening the interests (Jung, Chow & Wu, 2003). Through inspirational motivation leaders encourage and stimulate follower's creativity (Mittal & Dhar, 2015).

Creativity refers to the birth of new ideas for problem solving within organization (Mehraban & Avanloo, 2014). Transformational leaders provoke leaders to raise the expectation level and drag them to higher goals from self-oriented objectives (Gumusluoğlu & Ilsev, 2009). It is transformational leader's obligation to cater for subordinates needs so that they stick to their tasks (Jyoti & Dev, 2015). Organizational support provokes novelty (Mehraban & Avanloo, 2014) and pays back in different dimensions of resources as time, personal and pecuniary matters, directly affect creativity (DeConinck & Johnson, 2009). It is the push of transformational leaders to provide push for liberal ideas and abandoning old ways to treating the problem (Jassawalla & Sashittal, 2002). Realization of such support provokes creativity and innovation in the employees (Brockner & Higgins, 2001). At the stage, when they realize that organization is not orthodox and become liberal in accepting change, they become enthusiastic to take risks and actively work on new ideas without the fear of being punished (Brockner & Higgins, 2001)

In this modern world, every organization wants change and new efficient methods for handling problems. Organizational success can only be achieved by creative ways (Khalili, 2016). Almost every company tends to get work from employees but only focuses on the organizational innovation ignoring the individual's creativity and never considering creativity as important asset for company (Gumusluoglu & Ilsev, 2009). Real menace of the today's corporate sector is the ill direction of organizations and ignoring most imperative part that is employee's creativity (Gumusluoglu & Ilsev, 2009). Employee's creativity fosters on the tree of organizational support (Khalili, 2016).

Pakistan as a nation has developed and nurtured a lot in recent years which is being depicted in form of literacy rate that provides better way of business. Our industries are being blessed with rejuvenated engineers coming from high class universities, serving a catalyst for progress by using new skills, techniques and modern technology. Transactional approach is commonly practiced in Pakistan. Transaction leadership style is not encouraged creativity of employees, (Gumusluoglu & Ilsev, 2009). Transformational leadership acts as a precursor for employees' creativity (Shin & Zhou, 2003). Organizational environment is directly connected to the employees and leaders are responsible for evaluating, coaching and directing subordinates, and thus influence overall employees' creativity, well-being and performance (Epitropaki & Martin 2013; Battistelli & Mariani 2011). The organizational support and leadership style can boost employees' creativity at the workplace (Khalili, 2016).

In the past, many researchers (Khalili, 2016; Mittal & Dhar, 2015; Judge & Piccolo, 2004; Gumusluoglu & Ilsev, 2009) have done studies about organizational support, transformational leadership and employees' creativity separately. However, all of these studies lack attempts to collect and interpret the empirical research done by combining these topics. This study is conducted to investigate influence of transformational leadership and organizational support on employee's creativity.

There are the following pertinent research questions:

1. Does transformational leadership influence employee's creativity?
2. Does organizational support influence employee's creativity?

Pakistan has its own view of international management, which is being practiced in Pakistan. Formulated the paramount part of research that is the main research and the supporting questions and then we will throw light on objectives and aims of research.

The objective of this study is to:

1. To evaluate the influence of transformational leadership on employee's creativity

2. To evaluate the influence of organizational support on employee's creativity

Creativity is referred as ability to generate new and novel thoughts (Zhou & Shalley, 2003) in order to solve imminent and existing problems being faced by organizations or individuals (Mehraban & Avanloo, 2014). It can only be achieved by utterly facing the imminent problems with more novel solutions will and determination (Jyoti & Dev, 2015). So this study emphasize on enhancement of employees creative performance Transformational leadership became mesmerizingly popular over the years because of its unique and new methods to motivate and assist employees (Zhou & Shalley, 2003). Transformational leaders are directly associated with other dimensions that creates a productive work environment, which are motivation, confronting problems, clear vision, determination and elevating innovation and creativity (Mittal & Dhar, 2015). Organizational support is also considered as very import part in the encouragement of employee's creativity. This research emphasizes transformational leadership, organizational support and employee's creativity.

This research added knowledge in transformational leadership and organizational support and employee's creativity; it also added information about the relationship among transformational leadership, organizational support and employee's creativity. This research has emphasized to make inquiries on enhancement of employees creative performance. This research will now throw light on four elements of transformational leadership and tend to tell their effect on creative performance. This research will also enhance and elevate the knowledge and know-how of employees' novelty with assistance of transformational leaders and organizational support in telecommunication sector of Pakistan.

LITERATURE REVIEW

2.1 Employee Creativity

Creativity can be viewed as the generation of new, imaginative, slightly different and useful ideas (Zhou & Shalley, 2003). It is entirely the birth of individual's creative thought that leads to something very new and mostly it the result of individual's capacity, skill, learning, eagerness towards liberal thought and experience (Jassawalla & Sashittal, 2002). Change has become quite inevitable in our daily routine. Human resources are the most indispensable and valuable factor for running an organization effectively, efficiently and smoothly (Ali & Farid, 2016). This is necessity of every business. The accomplishment of a company mostly depends on the loyalty and hardworking of manages and employees. The supervision of employees is generally dependent on the quality of leadership (Bushra, Ahmad & Naveed, 2011).

The individual forms the basis for organizational innovation and he is the eventual source of every new idea (Shalley & Gilson, 2004). Therefore, hypothetically, the raw material essential for organizational novelty is provided by the employee's creative performance. Creative employees search and identify opportunities for launching new products. They can find new ways to use existing equipments or methods, or create unique, but realistic work-related ideas. Such kind of employees is more likely to consider creative clarifications to the hurdles and advocate ideas to others, but also formulate necessary plans for the implementation of their thoughts.

Creative employees come up with unique and useful ideas related to organizational practices, procedures or products (Shalley & Gilson, 2004). Also, these employees can influence others in the organization by offering a role model. Dewett (2006) suggest that new ideas presented by creative employees are transferable to others in the company for their own development and use.

Hence, such entity level creativity through formation and implementation is likely to result in the development of inimitable products at the organizational level (Dewett, 2006). Organizational progress and innovation are positively impacted by the creativity of its employees.

2.2 Transformational Leadership

"Leadership" is a practice by which an manager can express, direct, guide and manipulate the performance (Judge & Piccolo, 2004) and effort of others towards completion of specific goals in a given circumstances (Khalili, 2016). Leadership is the capacity of a manager to induce the subordinates to work with passion and confidence (Almutairi, 2016). During the last few years, transformational leadership a lot of research and focus has been paid because of long-lasting motivation (Piccolo & Colquitt, 2006). It is the leader's duty to craft an environment so that followers and subordinates can work in an environment that achieves the common goal (Almutairi, 2016). The transformational leaders transcend the common circle of people, they have a vision which overlaps the individual self-interests and focus more towards group, organization or society and are more visionary and dynamic (Piccolo & Colquitt, 2006). It is the main controller who drives the people towards the change needed and change group, teams by crafting and designing an environment to achieve intended missions and their activities greatly affect the result of business outcomes (Jyoti & Dev, 2015).

It is transformational leader's ability to influence the subordinates by elevating their confidence and craft missions for subordinates (Almutairi, 2016). It has four divisions: approaching first idealized influence/ charisma, leaders toil to develop sense of combined mission, homage, loyalty and motivation besides that it also has the influence to fancy their subordinates as role model (Almutairi, 2016). Leaders have moral obligations and adhere to code of conduct, which make them more respectable and develop a sense of likeliness; subsequently their subordinates tend to follow their leaders (Piccolo & Colquitt, 2006). Intellectual simulation is solely concerns with innovation and creation, which helps in mitigating the problem (Gong, Huang & Farh, 2009). Through inspirational motivation leaders encourage and stimulate followers (Jung, Chow & Wu, 2003). Individual consideration is imperative thing which assist the individual and make more contented. So, this subject is in spotlight (Piccolo & Colquitt, 2006).

2.3 Transformational Leadership and Employees Creativity

Creativity or novelty is a total blessing (Piccolo & Colquitt, 2006) which provides help in dual ways being assisted by organization environment without fear of being failure; one is the adaptation of new procedures for same old problems: and other is the business innovation that is creation of new product or service (Piccolo & Colquitt, 2006). Research finds relationship in transformational leadership and employees' creativity. There is also a very useful link that builds the trust of leaders, which enables the employees to overcome their fears and take risks needed to achieve organizational objectives (Piccolo & Colquitt, 2006). When the leaders begin to own the results of actions taken by their subordinates to resolve a problem by taking necessary risks, then it encourages them to work more and think creatively and independently (Dickson, Hartog & Mitchelson, 2003). Ability to produce unique ideas can bring by the leader through constructing innovative environment (Jong & Hartog, (2007).

Motivation is the backbone of creation which provokes subordinates to work in creative environment generated by leaders (Avolio & Bass, 1988). Assisting subordinates in problems provokes subordinates towards innovation. Supportive Supervision assist subordinates to work more innovatively (Piccolo & Colquitt, 2006). Transformational leaders are more helping in generating new ideas by subordinates and better solutions and ways to different problems (Bass & Riggio, 2006). Subordinates work with their maximum and work innovatively when they have complete belief and trust that can be provided by the leaders and can demand work beyond their expectation because they have vision of the organizational goals (Mittal & Dhar, 2015).

Transformational leaders who take subordinates together in order to gain the organizational goals, individual attention, motivate them with appreciation and team coordination (Mittal & Dhar, 2015). As compared to transactional style leadership, transformational leaders contribute more to the company's goal.

It has been found by different studies on leadership that a positive relationship is present between employee's creativity and leadership (Gong, Huang & Farh, 2009; Dickson, Hartog & Mitchelson, 2003). Critical thinking is stimulated through transformational leaders and they motivate their subordinates to take initiative and risk most of the time. They ensure support to their followers for taking risks and they own consequences of such risks and the actions of their employees. This support alters their behavior and motivates them to contribute into creative work processes (Dickson, Hartog & Mitchelson, 2003).

Transformational leadership are more of role models for their subordinates and driving force for elevating motivation and idealized/ charismatic increases novelty and referred creativity catalyst (Mittal & Dhar, 2015). Individualized consideration pay off acknowledgement and motivation, intellectual stimulation "increase" thought process by adopting liberal ideas and confronting problems; and inspirational motivation gives "encouragement" to the novelty generation and assists them to work for organization's mission (Gumusluoglu & Ilsev, 2009).Transformational leaders help in developing new designs for achievement of work and are ambitious to face new challenges (Jong & Hartog, 2007). The use of motivation and intellectual motivation enhances the creative ability of their followers (Mumford & Licuanan, 2004; Jung, Chow & Wu, 2003). Motivation urges subordinates to put more endeavors for solving problems and they willingly contribute in creative work behaviors. The positive relationship between transformational leadership and creativity of employees challenges them and energizes them to find unique approaches in their work (Hyman-Shurland, 2016).

Researchers have argued that the support offered by leaders to their followers in solving issues motivates them to engage in unique activities (Khalili, 2016; Mumford & Licuanan, 2004). (Jong & Hartog, 2007) found that subordinates engage more in creative actions when are properly supervised. Jung, Chow & Wu (2003) compared other forms of leadership with transformational leadership and stated that it is more effective at motivating followers to think critically and to adopt exploratory and generative thinking processes that yields more creative solutions and ideas. These above mentioned literature clearly prove that employees' creativity is enhanced by transformational leadership.

Hypothesis 1: Transformational Leadership is significantly associated with employee's creativity

There are four dimensions of Transformational leadership:

2.3.1 Idealized Influence

This part is very pertinent to transformational leaders transferring individual objectives to organizational goals and gives subordinates mission, trust, objectives, loyalty and vision (Bass & Riggio, 2006). It is the point which roves leaders to be more eager and determinant against problems and develops trust of employees which in return pay homage to their leaders and view them with respect, in return as part of this relation the subordinates get vision in appealing way (Mehraban & Avanloo, 2014). By leading with affective example and behaving confidently and risk taking with optimism enhance confidence of subordinate on their leader (Bass & Riggio, 2006). They deliver max of their output if they have belief in their leaders and these things assist them in creativity (Khalili, 2016). Transformational leaders surely act as a great source of inspiration and role model for employees. Such workforce has a strong faith in the supervision. The purpose of this form of leadership is to work hard for project accomplishment with confidence, motivation, creativity, trust, enthusiasm and risk bearing.

Hypothesis 1(a): Idealized influence is significantly associated with employee's creativity

2.3.2 Inspirational Motivation

This part deals with inspiration and motivation of employees towards new and specific goals and missions (Piccolo & Colquitt, 2006). Develop a meaning to their work by articulating a clear version of future visions and mission to them and provide elaborate mutual gains of both economic and mission ideologies (Khalili, 2016). Become role model and wore optimistic dress and try to stimulate group mission and try to overcome personal goals (Judge & Piccolo, 2004).

This will subordinate in creativity as they are free from internal misconceptions and trusting their leader in backing of failure (Judge & Piccolo, 2004). Provide with highly and excellent outcome of result and support gives them motivation and enthusiasm for achieving collective goals and mission articulated by their leaders (Piccolo & Colquitt, 2006).

Hypothesis 1(b): Inspirational Motivation is significantly associated with employee's creativity

2.3.3. Intellectual Stimulation

Intellectual Stimulation is a trait of leaders that they challenge their disciples' ideas and values for solution of problems. Transformational leader motivate followers in new way, make them creative and good decision maker (Keller, 2006). The transformational leader uses Intellectual Simulation technique in order to encourage subordinates towards creative thinking and innovation (Jyoti & Dev, 2015).

Confidence and faith can only be built by having a belief on subordinate's capabilities and allow him to adopt more novel ways regardless of fear of loss, which make them more critical and reasonable in their approach towards problem solving (Keller, 2006). Provide them enough of faith so that they can challenge status quo and thought freely of internal and external problems, which is lead of innovation and creativity most expected out of transformational leaders by the organizations (Jung, Chow & Wu, 2003).

Hypothesis 1(c): Intellectual stimulation is significantly associated with employee's creativity

2.3.4 Individual Consideration

It refers to handling cohorts as individuals and not just associates of a cluster. Such leader satisfies the individuals by supporting, advising and giving attention to their single needs and inspire them to improve themselves. This dimension of leadership enhances the capability of leader to pay more attention to followers (Hater & Bass, 1988).

This is chief part as it develops the employee's nurturing program and tries to lessen any concerns he or she is dealing with. In this type of leadership leader, enables every individual to contribute and share his ideas that can prove beneficial for the personal and organizational goals (Jyoti & Dev, 2015). It deals at individual level once given the desired collective goal (Bass & Riggio, 2006) and mission, much heed is paid to the betterment of employee and special deed to be done (Piccolo & Colquitt, 2006). Direct communication of leader with their subordinate boost the morale and give confidence to subordinate and afterwards recognition and acknowledgement parts are very important rather it be economic reward or simple appraisal words but it necessary (DeConinck & Johnson, 2009). If they are given discretionary power to complete the task in a given direction, they will put more effort and gave more time to work on their own (Judge & Piccolo, 2004). In addition to that private morale boosting can assist a lot and fair distribution of workload is important (Bass & Riggio, 2006). This way employee feels motivated and become more eager to learn and experience different problems and coming up with better solutions (DeConinck & Johnson, 2009).

Hypothesis 1(d): Individual consideration is significantly associated with employee's creativity

2.4 Organizational Support

Organizational support elevates novelty (DeConinck & Johnson, 2009) as well as gives heed, recognition and encouragement by assisting them in resources that are funding and time. Realization of such support provokes creativity and innovation in the employees (Brockner & Higgins, 2001). At the stage, when they realize that organization is not orthodox and become liberal in accepting change, they become enthusiastic to take risks and actively work on new ideas without the fear of being punished (Brockner & Higgins, 2001).

The term, organizational support, becomes viral in last few years and is still in research domain because it assists in understanding the relation between organization and subordinate point of view, which depict the employs behavioral outcome (Neves & Eisenberger, 2012). Organizational support is basically a measure of employees understanding regarding organizational values and contributions towards their well-being (Neves & Eisenberger, 2012). Briefly, it simply means the organizational attitude towards employees, what reaction against employee's illness, small and large mistakes, performances and awarding the salaries fairly against their work and motivation and try to assist employee in making tasks meaningful (Brockner & Higgins, 2001).

High organizational support makes employees happy, satisfied and more positive at work place (Brockner & Higgins, 2001). It enhances the trust which in return helps in gaining job commitment with the organization and elevated perceived organizational support (Neves &

Eisenberger, 2012). Very low perceived organizational support has instigated the negative domain activities like being late, less interested, little creativity, less motivation and absenteeism at work place (Brockner & Higgins, 2001).

2.5 Organizational Support and Employee's Creativity

According to (Shoghi, Asgarani & Ashnagohar, 2013) creativity is essential for a company's long-term development and survival. Employee's creativity as induced by organizational support to have competitiveness in aspiring globalized business environment (Mehraban & Avanloo, 2014). This analysis also asserts that creativity can only be achieved with support from the organization, however, Support comes in many forms including risk taking, developmental experiences, encouragement in relation to pecuniary and security support, provided by the organization for the employees (Brockner & Higgins, 2001). Organizational support is mostly considered to elevate employee's creativity and novelty (Mehraban & Avanloo, 2014).

Organizations can create actions and behaviors of creativity which are helpful to have an understanding of human resource management (Brockner & Higgins, 2001). Influence of organization behavior have different aspects of appraisal and motivation on the work environment which affects creative innovative work performance and provides an initiative to go for risk without the fear of being punished (Mehraban & Avanloo, 2014). In contemporary competitive business environment, providing organizational support to employees enables the organization to be successful (Brockner & Higgins, 2001). Employees appraised and supported by their organization often feel this support is given because they are more experienced and valuable employees for their organization (Brockner & Higgins, 2001).

As proved by many researches that creativity denotes a unique way to think about problem solution and bringing innovations in companies (Mehraban & Avanloo, 2014). (Salajeghe & Movagharzade, 2016) suggested that subordinate's creativity can be attained by organizational support. Numerous researchers have dedicated on the relationship between creativity and leadership style, as (Salajeghe & Movagharzade, 2016) analyzed the relationship between attitude of leaders and creativity.

Transformational leaders toil to provoke their employees that are mostly liberal in accepting change without caring for risks (DeConinck & Johnson, 2009). The creativity of employees is expected to come under study and become more important when the scale of measurement is innovation (Brockner & Higgins, 2001). It has impact on transformational leadership in two ways; first is the environment of organization, if they found surroundings more innovative then they tend to work more innovatively and become prone to more risks without any fear (DeConinck & Johnson, 2009); secondly, when they are provided with enormous support and enough resources, their response becomes more effective. So, in such a conducive environment, transformational leadership has stronger influence on employee's creativity. (Gumusluoğlu & Ilsev, 2009).

Hypothesis 2: Organizational support is significantly associated with employee's creativity

2.6 Transformational Leadership Theory

Transformational leadership is contemporary approaches that views on where the company should be destines and analyze how to administer the external and internal changes that followers need to achieve the goal. Today's real issue is the equilibrium between creativity and organization specific goals despite those transformational leaders are praise worthy of achieving desired outcome (Dewett, 2006). In order to make it different from other organizations, they have to craft an environment that is conducive for employee's innovation (Mittal & Dhar, 2015).The real edge comes to play when contribute to competitive edge for organization because that is the novel thing from routine practices and that is all origination's responsibility to assist them with an environment (Dewett, 2006).

The Bass Transformational Theory was interested in understanding to which extent leaders influence their followers. Employees follow their leader because of honesty, trust and other leadership qualities and their loyalty is directly dependent on how much these qualities are strong (Mittal & Dhar, 2015). Employee's transformation occurs due to these qualities.

Idealized influenced symbolizes to the supervisor becoming a completely developed role model, performing and signifying ideal traits of reliance, sincerity, pride and enthusiasm (Avolio, Bass & Jung, 1999). Inspirational motivation is provided by the leader and it gives unambiguous significance to the followers for a task (Avolio & Bass, 1995). This procedure involves providing an unambiguous vision or clear ambition. The leader will apply his charismatic approach in leading the team to go forward. A leader offers intellectual stimulation in the form of a challenge to the existing order, individual and task. The leader seeks new ideas from the team and encourages them to collaborate. He takes the role of a teacher and teaches them to learn and be independent. There is an emphasis on every team member needs and individual considerations (Avolio & Bass, 1995). He has to act as a mentor, teacher, role model or facilitator to bring the employee into the team and motivate them to do tasks.

2.7 Organizational Support Theory

Employees form a broad perception about the level to which the organization values their contributions and cares about their well-being (Eisenberger et al., 1986). Such perceived organizational support would increase employees' felt obligation to benefit the organization reach its objectives, their affective commitment to the organization, and their expectation that improved performance would be rewarded (Cotterell, Eisenberger & Speicher, 1992).

Organizational support theory is composed of the intention to understand the subordinates' reactions of the organization's positive processes (Eisenberger & Stinglhamber, 2011; Neves & Eisenberger, 2012). It gives the basis for understanding both origins and effects of organizational support/ For example, it explains the favorable actions taken by the employees such as their improved efforts. Employees represent their organization and give it humanlike traits, such as either benign or evil intentions towards them (Neves & Eisenberger, 2012).

Organizational support indicates valuation of subordinates' endeavors and motivation to offer support or help when requires as well as notifying to meet their socio-emotional needs. Generally a great part of this theory is explained by the trait of reciprocity because subordinates expect to get rewarded for their efforts and thus work that much harder. In short, this theory offers guidelines for the pragmatic implementation of advocating organizational support and its positive results (Eisenberger & Stinglhamber, 2011).

2.8 Theoretical Framework

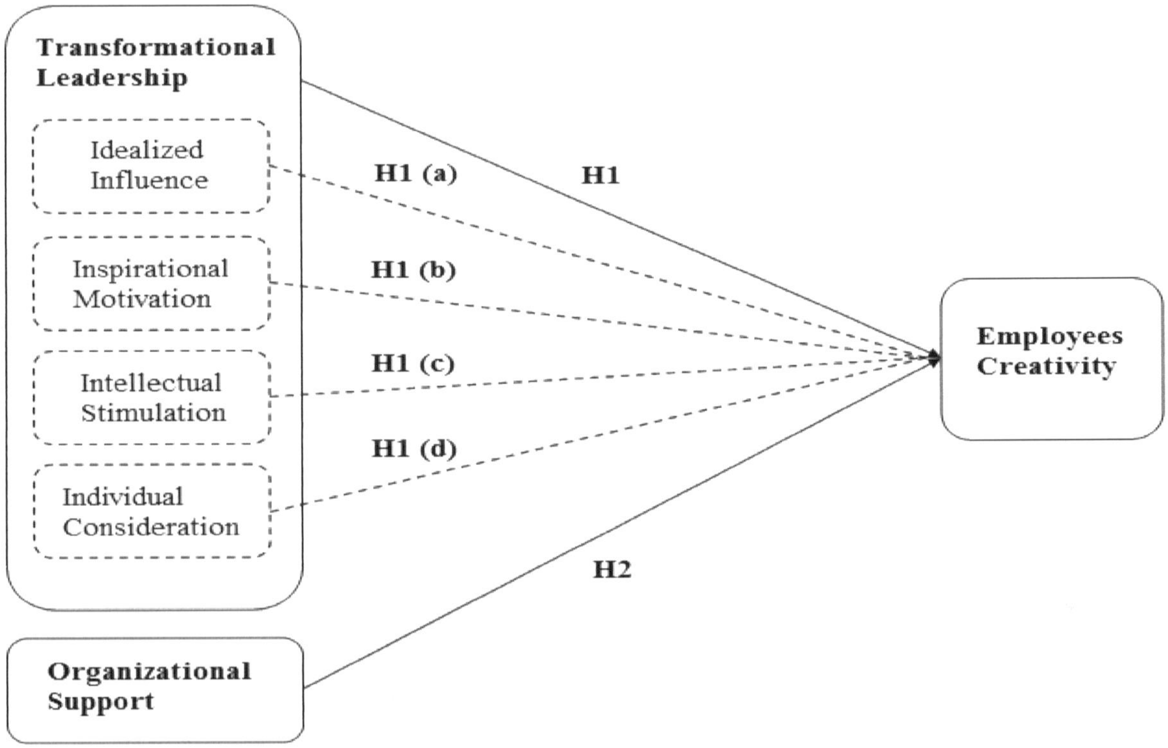

RESEARCH METHODOLOGY

This study has descriptive nature as it includes finding out the effects of transformational leadership on creativity of employees with higher management support in the telecommunication sector. The entire study centers about the significance of transformational leadership and organizational support with their effect on creativity of employees. Hypothesis based on the reality that various recognized elements play their role in accomplishing transformational leadership. These elements can be used to transmit leadership among the personnel of an organization. Those identified elements include: transformational leadership (idealized influence, inspirational motivation, intellectual stimulation and individualized consideration) and organizational support and their effects on creativity of employees.

The best method for collection of data is a structured questionnaire, especially for a cross sectional study. Thus, this study adopted this questionnaire approach for conducting study, which is most acceptable for collecting data from participants when a quantitative research is performed. From July 2016 to August 2016, 300 responses were received from the study participants but only 260 responses were valid. These were used for analyzing data. This study seeks to evaluate the relationship among the variables organizational support, transformational leadership and creativity of the employees. Transformational leadership and organizational support determines its own proportions and has diversified influence on creativity of the employee. The conducted research is exploratory in nature because the route and strength of relationship would be analyzed to recognize the concentration and type of the relationship.

The survey was divided into four sections. Demographic information comes first it's about Age, gender, qualification and organizational experience. The second section is related to transformational leadership information. The third section has organizational support information and the fourth section is related to employees' creativity. A self-administered questionnaire was used.

For conducting the research, such respondents had been selected who was serving in an organization and had ample experience and qualification. Hence, the study is performed by using the convenience sampling technique. The time constraint was another reason to use this technique. Also, this study was focused on organizations present in Islamabad.

Participants are selected from telecommunication sector. Telecommunication sector has seventeen categories i.e. Wireless, GSM, Radio etc. In Islamabad, so many networks have GSM. 14 companies in Islamabad have GSM. In year 2006-2007, total employees in telecommunication companies came out to be 84,000. (SIM activation Tax and Mobile

Telecommunication in Pakistan, a report by Deloitte 2007). Last figure for the total number of employees is not available, therefore it is assumed to be above 1 lakh. Hierarchy of the companies taken is the top management and middle management including project managers, General Managers and team members. Collection of data is done only from 2 companies Ufone and Mobilink so the population of selected companies is 800 so sample size is 260 according to Morgan's table. Participants are selected randomly. The participants are selected based on an entirely updated list of the population.

Employees are the most indispensable and valuable asset for running an organization. The accomplishment of a company mostly depends on the loyalty and hardworking of its employees. So for this study creative performance of employees was measured. So, the unit of analysis for this study is employee. Measurement scales used in this study are summarized in a Table 1.

FINDINGS AND DISCUSSIONS

The demographic information of the respondents is calculated, there were 260 respondents, their age, qualification and organizational experience is presented in Table 2.

There are correlation, regression, standard deviation and means of transformational leadership (inspirational motivation, idealized influence, individualized consideration, intellectual stimulation) and organizational support on employee's creativity.

According to reliability statistics the value of Cronbach's Alpha is within range and also given in Table 3.

The correlation of all components of transformational leadership is given Table 4 that presents the results of correlation among the components of transformational leadership. It has been observed by this study that there is a definite and significant correlation between employee's creativity and idealized influence because ($r=.562$ and $p< 0.01$) points out that there is a definite relationship between idealized influence and creativity of employees. Inspirational motivation has definite and significant relationship with creativity of employee because value of ($r=.701$ and $p< 0.01$) which indicates a significant and positive correlation with creativity of employees. Likewise, intellectual stimulation has definite and significant correlation with creativity of employees (because $r=.659$ and $p<0.01$) which indicates correlation is significant at 0.01 level. The last one, individualized consideration also has definite and positive correlation with creativity of employee because ($r=0.659$ and $p<0.01$) which indicates correlation is significant. Above given table presents the results of correlation among all independent and dependent variables of the study. The collective effect of all the components of transformational leadership is significant with creativity of employees because ($r=0.765$ and $p<0.01$) that demonstrates correlation is significant at 0.01 level. Likewise, organizational is significant with creativity of employees because ($r=.715$ and $p< 0.01$) which demonstrates at 0.01 level correlation is significant.

The regression analysis is given under below Table 5. The study has done regression analysis of all variables of transformational leadership. These variables are not only discussed separately but also a combined effect of all of them into one variable which is transformational leadership is examined with creativity of employees to assess the combine effect. The predicted regression analysis of transformational leadership and creativity of employee is statistically significant. "It is noticed that creativity of employees is strongly impacted by the transformational leadership. It can be examined that the f-statistic = 365.171 is significant at $p < 0.05$ which indicated that transformational leadership clearly explains the variation in creativity of employee. R-square shows 0.586, $p < 0.01$ proportion of variation in the dependent variable explained by the regression model".

The predicted regression analysis the creativity of employee and organizational support is also statistically significant. It can be examined that f-statistic = 307.742 is statistically significant at $p < 0.05$ indicating transformational leadership clearly explaining the difference in creativity of employee. As the results of R-square shows 0.544, $p < 0.01$ proportion of variation in the dependent variable described by regression model.

Results and Discussions

The tests performed on the hypothesis are separately discussed as under:

Study has done statistical analysis of all variables of transformational leadership. These variables are not only discussed separately but also has a combined effect of all of them into one variable which is transformational leadership is examined with creativity of employees to assess the combine effect of the intended study. Hypothesis 1 is associated with all the components of transformational leadership suggested formerly have been discussed below and proven. Thus transformational leadership is significantly associated with employee's creativity, i.e. ($r = .765$, $p<0.01$). The standardized beta value ($\beta = 0.765$, $p < 0.01$) and f-statistic value ($\Delta F = 365.17$, $p < 0.01$) indicates significant relationship between employees creativity and transformational leadership.

Research assists the primitive study (Gong, Huang & Farh, 2009; DeConinck & Johnson, 2009) that descript a significant association between transformational leadership and employee's creativity. Literature is asserting that there is positive association between transformational leadership and employee's creativity. When leaders adopt the role transformational leadership, automatically employee's novelty increase and it is the paramount element for creativity and novelty of employees (Shin & Zhou, 2003). Basically it is attached to environment that automatically encourages creativity (Sendjaya, Sarros & Santora, 2008). It is supported by results that transformational leadership and employee's novelty is significantly associated with each other.

The formerly stated correlation results show that there is a significant association amongst idealized influence and employees creativity, i.e. ($r = .562$, $p<0.01$). Additionally idealized influenced has examined from regression analysis with beta value ($\beta = .562$, $p < 0.01$) and f-statistic value ($\Delta F = 119.10$, $p < 0.01$) indicates significant relationship between idealized influence and employees creativity. In the past, many researchers (Shin & Zhou, 2003; Khalili, 2016) have studied the significance of the idealized influence on employee's creativity. Idealized influence is focused on generating condition where employees can perform at the optimal level (Khalili, 2016). Through results of this study it is proved that this part is very relevant for transformational leader to boost the employee's creativity by changing individual objectives to organizational objectives and give employees trust, objective, mission, vision and loyalty.

The mentioned correlation results show that there is a significant association between inspirational motivation and employees creativity, i.e. ($r = .701$, $p<0.01$). Additionally inspirational motivation has examined from regression analysis with beta value ($\beta = .701$, $p < 0.01$) and f-statistic value ($\Delta F = 248.93$, $p < 0.01$) indicates significant relationship between idealized influence and employees creativity. This clearly shows that the impact of inspirational motivation on employee's creativity is not only definite but studies ((Piccolo & Colquitt, 2006; Judge & Piccolo, 2004) also confirm that it also enhances commitment and satisfaction towards the organization. It can be viewed from the conducted research that inspirational motivation is more expected to boost job involvement, creative performance and enhances motivation to learn novel ways of creativity.

The previously mentioned correlation results show that there is a significant association between intellectual stimulation and employees creativity, i.e. ($r = .659$, $p < 0.01$). Additionally intellectual stimulation has examined from regression analysis with beta value ($\beta = .659$, $p < 0.01$) and f-statistic value ($\Delta F = 197.75$, $p < 0.01$) indicates significant relationship between intellectual stimulation and employees creativity. Studies ((Bass & Riggio, 2006; Keller, 2006)) confirmed that transformational leaders stimulate the employees in a unique way and boost creativity by giving them confidence of making timely and risky decision. It requires liberal solutions to old problems and new procedures and ways in more logical and reasonable manner.

The previously mentioned correlation results show that there is a significant association between individual consideration and employees creativity, i.e. ($r = .725$, $p < 0.01$). Additionally individual consideration has examined from regression analysis with beta value ($\beta = .725$, $p < 0.01$) and f-statistic value ($\Delta F = 286.262$, $p < 0.01$) indicates significant relationship between intellectual stimulation and employees creativity. Studies (Bass & Riggio, 2006; Piccolo & Colquitt, 2006) confirmed that individual consideration denotes the degree, to which any leader caters to each follower's individual needs, acts as a mentor or teach the follower, and listens to his follower's concerns. Individual consideration is concerned with individual development by enhancing their self-esteem and getting rid of fear of failure through cultivating confidence. Thus it is confirmed that individual consideration boosts creative abilities of individual.

The formerly mentioned correlation results show that there is a significant association between organizational support and employees creativity, i.e. ($r = .715$, $p < 0.01$). Additionally organizational support has examined from regression analysis with beta value ($\beta = .715$, $p < 0.01$) and f-statistic value ($\Delta F = 307.74$, $p < 0.01$) indicates significant relationship between organizational support and employees creativity. Studies (DeConinck & Johnson, 2009; Brockner & Higgins, 2001) confirmed that it is predominant attribute of organizational support to increase uniqueness as well as giving attention, encouragement and recognition by aiding them in resources that are funding. Realization of this support generates innovation and creativity in the employees.

CONCLUSION

With the commencement of new modern era every firm has to evolve with the requirements. It has been observed that novel employees and transformational leaders assist in creating firms more novel and more creative. Creativity environment is the most paramount thing in the success of organization and in ascending employee's novelty. There is plethora of search being conducted between creativity and transformational leadership however mainly its relation to motivation, to satisfaction as well as to performance and other factors but contemporary study is pertinent to transformational leadership and organizational support on employee's novelty and creativity accompanied by organizational support. Organizations stuffed with creative employees tend to enjoy competitive edge and develops additional worth in healthy business environment. Constraints of work environment as leadership style immensely increase the creativity in employees.

In this modern era telecom sector need to be update over time and the competitive advantage with innovation comes with resources, technology and capabilities. Company should activate its resources of capabilities to achieve competitive position in market. It is innovation that leads the company to resourcefulness and lofty valuable creativity which is source of competitive advantage. Support of organization motivates and alleviates the urge of innovation in employee and in other words organizational support enhances the employee's creativity. This study describes that transformational leadership is prerequisite of employee's creativity. Leaders are mentors and are responsible for evaluation, directing and coaching subordinates and altering the behavior, needs and employee's creativity. Personality of leader on one shoulder and organizational support on another shoulder helps in enhancing creativity.

Practical Implications

Aspiring environmental and growth sustainability of businesses is dependent largely on employee's creative ideas and innovative solutions to environmental problems. This particular study analyzes how firms that intend to improve their environmental and growth performance can significantly increase their employee's willingness to innovation and risk taking initiative. Result deduced from an employee survey shows that supportive and encouraging behaviors from line managers and high ups significantly ascend the probability that employees will try innovation and environmental initiatives to come up with new ideas to counter the contemporary challenges. Even though in firms that have sustainable and enduring development policies and rules, line managers are less supportive and cooperative when managing and adhering environmental and business issues compared to dealing and managing other business problems.

Business and environmental policies may not directly influence the employee's willingness to innovation, creativity and risk involvement. This analysis of employee's creativity shows how companies can and some do, encourage their line managers to inculcate those behaviors that employees perceive as encouraging and supportive role and actions of firm, however, it indicates little appraisal from firms which appearing to be encouraging can proved very vital for aspiring business cooperation.

It is mandatory that line managers and high up managers pay consideration to employees' needs and a necessity which mostly includes security, monetary or nonmonetary and appreciation are the main demands of every employee. For this very reason managers and high ups, must have their focus on claims arising from the postponed salary and also offering promotion to the deserving employees. However, managers should have a plan to privilege everyone according to the needs and toiled to mitigate delayed payments. Creating an environment which employees achieve career promotion can be helpful to enhancing job performance as well as a source of motivation for employees to take steps towards creativity and risk involvement without the fear of punishment.

It is entirely leader responsibility to find the way out for innovation considering in mind the importance of employee's creativity in organization's success. Leaders should depict the transformational attitude and should provide the courage to go for novelty, share the vision, give them motivation, providing intellectual support and toiling to fulfil the needs of employees. Promotion is equally attracting by doing so you can also motivate your employees. However, leadership seems to be more tilt towards the regulatory focus and its imperative nature to provide the monitory context for promotion. It is firms' job to draft an environment filled with endorsement of success and novelty avoiding failures.

Limitations and Future Research Directions

This study describes significance of transformational leadership and organizational support on employee's creative performance. Telecom technologies are one of the dynamic technologies in world. Their product life cycle is very short, so they need fast thinkers and transformational leaders to cope up with changes. So it suggests companies to promote and encourage transformational leadership. Transformational leadership performs an important role in assisting employees as well as organizations to renew, create and innovate.

It is suggested that future study should be done along the longitudinal data and the innovation at organization level is very complex business being affected by the organizational glitches, environmental and personal affairs. It is longitudinal measurement that gives precision and gives its results systematically. Many constraints such as age, education, gender and experience should be captured however; some additional variables can be added in coming studies for example individual effort, group effort and psychological job. This study experiences the transformational leadership affecting the creativity along the road of organizational support. This study should be carried in other cultures to experience the effect of transformational leadership so that it can be verified. Future studies could extend this research by considering multi-dimensional top management support as a potential moderator or mediator variable, including different areas, such as health, education, manufacturing, services, and information technology.

There are constraints which will be conducive for future work as the respondents were the inhabitants of Islamabad so it can be asserted that this cannot be the general structure of whole Pakistan. It was time and money constraint that hampered and restricted Islamabad. Second big constraint is the cross-sectional form of study where researchers can study specific scenario at the time of survey in organization. Data was collected only from telecommunication sector.

6. References

Ahmad, F., Abbas, T., Latif, S., & Rasheed, A. (2014). Impact of Transformational Leadership on Employee Motivation in Telecommunication Sector. *Journal of Management, 2*(2), 11-25.

Allix, N. M. (2000). Transformational Leadership Democratic or Despotic? *Educational Management & Administration, 28*(1), 7-20.

Almutairi, D. O. (2016). The Mediating Effects of Organizational Commitment on the Relationship between Transformational Leadership Style and Job Performance. *International Journal of Business and Management, 11*(1), 231.

Amabile, T. M., Barsade, S. G., Mueller, J. S., & Staw, B. M. (2005). Affect and creativity at work. *Administrative science quarterly, 50*(3), 367-403.

Avolio, B. J., & Bass, B. M. (1988). Transformational leadership, charisma, and beyond.

Avolio, B. J., & Bass, B. M. (1995). Individual consideration viewed at multiple levels of analysis: A multi-level framework for examining the diffusion of transformational *leadership. The Leadership Quarterly, 6*(2), 199-218.

Avolio, B. J., & Bass, B. M. (1995). Individual consideration viewed at multiple levels of analysis: A multi-level framework for examining the diffusion of transformational leadership. *The Leadership Quarterly, 6*(2), 199-218.

Avolio, B. J., Bass, B. M., & Jung, D. I. (1999). Re□examining the components of transformational and transactional leadership using the Multifactor Leadership. *Journal of occupational and organizational psychology, 72*(4), 441-462.

Ali, S., & Farid, F. (2016). Effect of Transformational Leadership on Job Satisfaction and Organizational Commitment. *Available at SSRN 2713386.*

Bass, B. M., & Riggio, R. E. (2006). Transformational leadership. Psychology Press.

Battistelli, A., & Mariani, M. G. (2011). Perceived Organizational Support: Validation of the Italian version of the Survey of Perceived Organizational Support (with eight items). *Giornale italiano di psicologia, 38*(1), 189-214.

Brockner, J., & Higgins, E. T. (2001). Regulatory focus theory: Implications for the study of emotions at work. *Organizational behavior and human decision processes, 86*(1), 35-66.

Bushra, F., Ahmad, U., & Naveed, A. (2011). Effect of transformational leadership on employees' job satisfaction and organizational commitment in banking sector of Lahore (Pakistan). *International journal of Business and Social science, 2*(18).

De Jong, J. P., & Den Hartog, D. N. (2007). How leaders influence employees' innovative behaviour. *European Journal of innovation management, 10*(1), 41-64.

DeConinck, J. B., & Johnson, J. T. (2009). The effects of perceived supervisor support, perceived organizational support, and organizational justice on turnover among salespeople. *Journal of Personal Selling & Sales Management, 29*(4), 333-350.

Dewett, T. (2006). Exploring the role of risk in employee creativity. *The Journal of Creative Behavior, 40*(1), 27-45.

Dickson, M. W., Den Hartog, D. N., & Mitchelson, J. K. (2003). Research on leadership in a cross-cultural context: Making progress, and raising new questions. *The leadership quarterly, 14*(6), 729-768.

Eisenberger, R., & Stinglhamber, F. (2011). Behavioral outcomes of perceived organizational support.

Eisenberger, R., & Stinglhamber, F. (2011). Perceived organizational support: Fostering enthusiastic and productive employees. *American Psychological Association.*

Eisenberger, R., Stinglhamber, F., Vandenberghe, C., Sucharski, I. L., & Rhoades, L. (2002). Perceived supervisor support: contributions to perceived organizational support and employee retention. *Journal of applied psychology, 87*(3), 565.

Elkins, T., & Keller, R. T. (2003). Leadership in research and development organizations: A literature review and conceptual framework. *The Leadership Quarterly, 14*(4), 587-606.

Elkins, T., & Keller, R. T. (2003). Leadership in research and development organizations: A literature review and conceptual framework. *The Leadership Quarterly, 14*(4), 587-606.

Epitropaki, O., & Martin, R. (2013). Transformational–transactional leadership and upward influence: The role of relative leader–member exchanges (RLMX) and perceived organizational support (POS). *The Leadership Quarterly, 24*(2), 299-315.

García-Morales, V. J., Lloréns-Montes, F. J., & Verdú□Jover, A. J. (2008). The effects of transformational leadership on organizational performance through knowledge and innovation. *British journal of management, 19*(4), 299-319.

George, J. M., & Zhou, J. (2001). When openness to experience and conscientiousness are related to creative behavior: an interactional approach. *Journal of applied psychology, 86*(3), 513.

Gilson, L. L., & Shalley, C. E. (2004). A litTransformational leadershipe creativity goes a long way: An examination of teams' engagement in creative processes. *Journal of management, 30*(4), 453-470.

Gong, Y., Huang, J. C., & Farh, J. L. (2009). Employee learning orientation, transformational leadership, and employee creativity: The mediating role of employee creative self-efficacy. *Academy of management Journal,52*(4), 765-778.

Gumusluoglu, L., & Ilsev, A. (2009). Transformational leadership, creativity, and organizational innovation. *Journal of business research, 62*(4), 461-473.

Gumusluoğlu, L., &Ilsev, A. (2009). Transformational Leadership and Organizational Innovation: The Roles of Internal and External Support for Innovation. *Journal of Product Innovation Management, 26*(3), 264-277

Hourcade, D., Holers, V. M., & Atkinson, J. P. (1989). The regulators of complement activation (RCA) gene cluster. *Advances in immunology, 45*, 381-416

Hyman-Shurland, Y. (2016). The Merits of Trust in Transformational Leadership.

Jassawalla, A. R., &Sashittal, H. C. (2002). Cultures that support product-innovation processes. *The Academy of Management Executive, 16*(3), 42-54

Judge, T. A., & Piccolo, R. F. (2004). Transformational and transactional leadership: a meta-analytic test of their relative validity. *Journal of applied psychology, 89*(5), 755.

Jung, D. I., Chow, C., & Wu, A. (2003). The role of transformational leadership in enhancing organizational innovation: Hypotheses and some preliminary findings. *The Leadership Quarterly, 14*(4), 525-544.

Jyoti, J., & Dev, M. (2015). The impact of transformational leadership on employee creativity: the role of learning orientation. *Journal of Asia Business Studies, 9*(1), 78-98.

Kark, R., & Van Dijk, D. (2007). Motivation to lead, motivation to follow: The role of the self-regulatory focus in leadership processes. *Academy of Management Review, 32*(2), 500-528.

Keller, R. T. (2006). Transformational leadership, initiating structure, and substitutes for leadership: a longitudinal study of research and development project team performance. *Journal of applied psychology,91*(1), 202.

Khalili, A. (2016). Linking transformational leadership, creativity, innovation, and innovation-supportive climate. *Management Decision,54*(9).

Khan, M. M., Sarwar, A., Malik, S. A., & Ahmad, M. (2014). Influence of Transformational Leadership on Organizational Innovation in Telecommunication Industry in Pakistan. *Asian Journal of Business Management, 6*(3), 138-145.

Limbare, S. (2012). Leadership styles & conflict management styles of executives. *Indian Journal of Industrial Relations*, 172-180.

Masood, S. A., Dani, S. S., Burns, N. D., & Backhouse, C. J. (2006). Transformational leadership and organizational culture: the situational strength perspective. Proceedings of the Institution of Mechanical Engineers, Part B: *Journal of Engineering Manufacture, 220*(6), 941-949.

McGarry, J., & O'leary, B. (2013). The politics of ethnic conflict regulation: Case studies of protracted ethnic conflicts. *Routledge.*

Mehraban, M., & Avanloo, A. A. (2014). The Effect of Organizational Culture on Employees' Creativity. Arth prabandh: *A Journal of Economics and Management, 3*(8), 111-126.

Mittal, S., &Dhar, R. L. (2015). Transformational leadership and employee creativity: mediating role of creative self-efficacy and moderating role of knowledge sharing. *Management Decision, 53*(5), 894-910.

Mumford, M. D., & Licuanan, B. (2004). Leading for innovation: Conclusions, issues, and directions. *The leadership quarterly, 15*(1), 163-171.

Neves, P., & Eisenberger, R. (2012). Management communication and employee performance: The contribution of perceived organizational support. *Human Performance, 25*(5), 452-464

Ofori, G. (2009). Ethical leadership: Examining the relationships with full range leadership model, employee outcomes, and organizational culture. *Journal of Business Ethics, 90*(4), 533-547.

Piccolo, R. F., & Colquitt, J. A. (2006). Transformational leadership and job behaviors: The mediating role of core job characteristics. *Academy of Management journal, 49*(2), 327-340.

Pieterse, A. N., Van Knippenberg, D., Schippers, M., & Stam, D. (2010). Transformational and transactional leadership and innovative behavior: The moderating role of psychological empowerment. *Journal of Organizational Behavior, 31*(4), 609-623.

Salajeghe, S., & Movagharzade, A. (2016). Explaining the relationship between organizational commitment, organizational health and employee creativity of the Social Security Organization. *International Journal of Humanities and Cultural Studies (IJHCS) ISSN 2356-5926, 2*(4), 1010-1026.

Sarros, J. C., Cooper, B. K., & Santora, J. C. (2008). Building a climate for innovation through transformational leadership and organizational culture. *Journal of Leadership & Organizational Studies, 15*(2), 145-158.

Sendjaya, S., Sarros, J. C., & Santora, J. C. (2008). Defining and measuring servant leadership behaviour in organizations. *Journal of Management studies, 45*(2), 402-424.

Shalley, C. E., & Gilson, L. L. (2004). What leaders need to know: A review of social and contextual factors that can foster or hinder creativity. *The Leadership Quarterly, 15*(1), 33-53.

Shibru, B., & Darshan, G. M. (2011). Effects of transformational leadership on subordinate job satisfaction in leather companies in Ethiopia.*International Journal of Business Management & Economic Research,3*(5).

Shin, S. J., & Zhou, J. (2003). Transformational leadership, conservation, and creativity: Evidence from Korea. *Academy of management Journal, 46*(6), 703-714.

Shin, S. J., & Zhou, J. (2007). When is educational specialization heterogeneity related to creativity in research and development teams? Transformational leadership as a moderator. *Journal of applied Psychology, 92*(6), 1709.

Shoghi, B., Asgarani, M., & Ashnagohar, N. (2013). Mediating Effect of Organizational structure on the Relationship between Managers' Leadership Style and Employees' Creativity (Case study: Metal Industries of Kaveh Industrial City). *International Journal of Learning and Development, 3*(3), 63-80.

Shoss, M. K., Eisenberger, R., Restubog, S. L. D., & Zagenczyk, T. J. (2013). Blaming the organization for abusive supervision: The roles of perceived organizational support and supervisor's organizational embodiment. *Journal of Applied Psychology, 98*(1), 158.

Silverthorne, C. (2004). The impact of organizational culture and person-organization fit on organizational commitment and job satisfaction in Taiwan. *Leadership & Organization Development Journal, 25*(7), 592-599

Wang, H., Law, K. S., Hackett, R. D., Wang, D., & Chen, Z. X. (2005). Leader-member exchange as a mediator of the relationship between transformational leadership and followers' performance and organizational citizenship behavior. *Academy of management Journal,48*(3), 420-432

Zhou, J., & Shalley, C. E. (2003). Research on employee creativity: A critical review and directions for future research. *Research in personnel and human resources management, 22*, 165-218.

Table 1: Measurement of Variables

Variables	No. of Items	Adapted from	Scale
Transformational Leadership	12	Bass & Avolio, 1997	5-Point Likert
1. Idealized Influence	3		
2. Inspirational Motivation	3		
3. Intellectual Stimulation	3		
4. Individual Consideration	3		
Organizational Support	8	DeConinck & Johnson, 2009	5-Point Likert
Employees Creativity	9	Zhou & George, 2001	5-Point Likert

Table 2: Summary of demographic information

Item		N
Gender	Male	188
	Female	72
	Total	260
Age	20-25	44
	26-35	99
	36-Above	117
Qualification	Diploma Holder	0
	Bachelors	0
	Masters	86
	M.Phill�/MS	174
	PHD	0
Organizational-Experience	1-3	42

4-6	87
7-9	131
10-Above	0

Table 3: Reliability

Variables	Items		N0. of Items	Cronbach's Alpha
Transformational Leadership	TL_1	.805	12	0.882
	TL_2	.807		
	TL_3	.761		
	TL_4	.837		
	TL_5	.767		
	TL_6	.730		
	TL_7	.837		
	TL_8	.766		
	TL_9	.843		
	TL_10	.795		
	TL_11	.810		
	TL_12	.807		
Organizational Support	OS_1	.762	8	0.795
	OS_2	.719		
	OS_3	.843		
	OS_4	.656		
	OS_5	.651		
	OS_6	.729		
	OS_7	.805		
	OS_8	.726		
Employees Creativity	EC_1	.636	9	0.767
	EC_2	.719		
	EC_3	.752		
	EC_4	.680		
	EC_5	.636		
	EC_6	.680		
	EC_7	.581		
	EC_8	.775		
	EC_9	.581		

Table 4: Correlation Analysis

Sr.	Variables	1	2	3	4	5	6	7
1	Idealized Influence	1						
2	Inspirational Motivation	.475**	1					

3	Intellectual Stimulation	.374**	.346**	1				
4	Individual Consideration	.598**	.658**	.374**	1			
5	Transformational leadership	.567**	.660**	.524**	.585**	1		
6	Organizational support	.591**	.736**	.319**	.601**	.948**	1	
7	Employees Creativity	.562**	.701**	.659**	.725**	.765**	.715**	1

Table 5: Results of Separate Regression Analyses

Sr.	Variable	Model Summary			ANOVA		Coefficients	
		R	R^2	Adj R^2	f	Sig.	t	Sig.
1	Idealized Influence	.562[a]	.316	.313	119.105	.000[b]	10.914	.000
2	Inspirational Motivation	.701[a]	.491	.489	248.936	.000[b]	15.778	.000
3	Intellectual Stimulation	.659[a]	.434	.432	197.750	.000[b]	14.062	.000
4	Individual Consideration	.725	.526	.524	286.262	.000[b]	16.919	.000
5	Transformational Leadership	.765	.586	.584	365.171	.000[b]	19.109	.000
6	Organizational Support	.738	.544	.542	307.742	.000[b]	17.543	.000

Questionnaire

This survey questionnaire is designed to determine "Impact of transformational leadership and organizational support on employee's creativity". This Questionnaire is divided into four sections, first section relates to some basic demographic information, Second section relates to transformational leadership information, third section relates to organization support information and fourth section relates to employees creativity * Required

Gender *

- Male
- Female

What is your age? *

- 20-25
- 26-35
- 36-Above

What is your qualification? *

- Diploma holder
- Bachelors
- Masters
- MS/M.Phil
- PHD

Organizational Experience *

- 1-3 years
- 4-6 years
- 7-9 years
- 10- Above

These are the questions to find the strength of transformational leadership

1. Strongly Disagree
2. Disagree

3. Unsure
4. Agree
5. Strongly Agree

	Transformational Leadership	1	2	3	4	5
1	Is viewed as a symbol of success and accomplishment (CH) *					
2	Has a sense of mission which he/she communicates to me (CH) *					
3	Makes me feel good when I'm around him/her (CH) *					
4	Has a vision that spurs me on (IM) *					
5	Uses symbols and images to focus our efforts (IM) *					
6	Communicates expectations of high performance to me (IM) *					
7	Enables me to think about old problems in new ways (IS) *					
8	Emphasizes the use of intelligence to overcome obstacles (IS) *					
9	Requires that I back up my opinions with good reasoning (IS) *					
10	Provides advice to me when I need it (IC) *					
11	Gives newcomers a lot of help (IC) *					
12	Gives personal attention to those who seem neglected (IC) *					

The abbreviations in parentheses refer to: CH (Charisma), IM (Inspirational Motivation), IS (Intellectual Stimulation), IC (Individualized Consideration)

These are the questions to find out to what extent your organization supports you

1. Strongly Disagree
2. Disagree
3. Unsure

4. Agree
5. Strongly Agree

	Organizational Support	1	2	3	4	5
1	My organization strongly considers my goals and values *					
2	My organization really cares about my well-being *					
3	My organization would understand if I were unable to finish a task on time *					
4	My organization cares about my general satisfaction at work *					
5	My organization cares about my opinions *					
6	My organization tries to make my job as interesting as possible *					
7	My organization takes pride in my accomplishments at work *					
8	My organization is willing to extend itself in order to help me perform my job to the best of my ability *					

How employees go about accomplishing the creativity act

1. Strongly Disagree
2. Disagree
3. Unsure
4. Agree
5. Strongly Agree

	Employees creativity	1	2	3	4	5
1	Creative ideas simply occur to me without even thinking about them *					
2	I often use the technique of brainstorming to come up with new ideas *					
3	When I get a new idea, I get totally absorbed by it until I have pursued it completely *					
4	When I am generating new ideas, I do not tend to evaluate them until I have generated by ideas *					
5	I am at my creative best when I work alone *					
6	I am at my creative best when I work one other person *					

7	I am at my creative best when I work in a group *					
8	I typically create new ideas by combining existing ideas *					
9	I tend to work on my ideas simultaneously *					

DETERMINANTS OF KAIZEN AND KANBAN TECHNIQUES' ADOPTION IN THE INDUSTRIAL ESTATE OF RUSAYL, SULTANATE OF OMAN

Dileep Kumar M.
Professor
Berjaya University College
Malaysia. 55100.
dileep.kumar@berjaya.edu.my

Brinda Kalyani. P R
Assistant Professor
College of Education, Rustaq
P B No:10, P C 329, Sultanate of Oman
sradhaarjun@gmail.com

Abstract: Though different methods for costing are available to the world, Kaizen Costing and Kanban Technique are two methods, having its source from Japan which have secured world attention in its application and efficiency for inventory management and Lean Manufacturing. Oman is one of the fast developing countries and their industrialized communities have adapted this accounting technique for continuous improvement in a volatile business economy. A study was conducted with a macro research objective in order to explore the efficiency of Kaizen Costing and Kanban Technique, in the Rusayl Industrial Estate, of Sultanate of Oman. The study followed interpretive research philosophy with qualitative research methodology. Triangulated tools like interview, observation and document analysis were used to explore and understand the phenomenon under study through case study methodology.

Keywords: Kanban, Kaizen, Inventory Management. Supplier Interest, Quality Control, Management Commitment,

Reference: Kumar, D.M.& Kalyani, B. (2017). 'Determinants of Kaizen and Kanban Techniques' adoption in the Industrial Estate of Rusayl, Sultanate of Oman.'

INTRODUCTION

Kaizen Costing and Kanban Technique are two methods which are of Japanese origin but used worldwide as the best techniques for inventory management and Lean Manufacturing. Kaizen costing means continuous costing which will surely bring down the cost of production, which is tested for time and again to bring down the cost when we aim for target costing. Authors found Rusayl Industrial Estate to be the best area to be explored because it is the back bone of the economy next to PDO (Petroleum Development Oman). The respondents are the accountants of manufacturing companies who explained the techniques' applicability and suitability. There is total population of 230 enterprises in the Rusayl Industrial Estate, out of which 46 enterprises' accountants are taken as sample size. The data collected were analyzed on the basis of partial least square analysis and provide better insight into the adoption intention of Kaizen Costing and Kanban Technique in industries of Sultanate of Oman.

REVIEW OF LITERATURE

Kaizen is the process application for continuous improvement. As a long term competitive strategy, Kaizen is considered as one of the inevitable practice that followed by several multinational company. While, Kanban is one of the method used by industries to achieve Just in time in their production as well as allied process. The relationship between these two processes for continuous improvement is very much unique. It will ensure the organization improves at a more rapid rate. Huda (1992) found that Kaizen is practice that can be applied beyond the cultural differences for continuous improvement. The entire approach is people and the process oriented; hence it ensures the result efficiently. As it is evidenced a group of employees with inappropriate skills directly related to production contribution, their competence and participation in decision making can thus result into better output (Burch (2008). There are studies which supports the relationship between cost components and the profitability (Jayeola, (2012). Industrial health depends on identified cost Management techniques. Knowledge of Kaizen or Kanban won't permit the company to get it implemented unless and until the top management is committed. Further the industry should facilitate better vendor participation, and inventory management for Kanban deployment (Rahmana, 2013)

PROBLEM FORMULATION

Sultanate of Oman like any other GCC countries is undergoing a critical phase in economy because of the oil price plunge. The inflation report released on August 2017 reviews the most inflation rates during the month, and prices movement in the various provinces. The report reviews the data in a contemporary flair combining statistical tables, charts and the use of geographic information systems. The inflation rate in Oman was recorded at 1 percent in August of 2017. Since oil prices have gone down and also inflation runs in the economy it is necessary to go by Target Revenue minus Target Profit equals Target Cost where the business house need to identify the cause of the cost. If it keeps on cutting the cost without identifying the cause of the cost, the cost will be caused again and again. It says our management is poor. So target costing at this juncture gains a lot of attention. Target costing can be achieved through techniques of Kanban and Kaizen which was shown by Japanese experiments. However the implementation or the adoption of Kaizen and Kanban techniques depends on several factors like proper inventory management, supplier participation, quality control and quality improvement effort and top management support. So far fewer studies have oriented to explore the factors correlated to Kaizen and Kanban techniques in the Omani context. Therefore the authors have conducted a quantitative study on the adoption intention of Kaizen and Kanban techniques since it would be of great support to control the costs and thereby bringing forth a steady profit to business organizations.

THEORETICAL FRAMEWORK

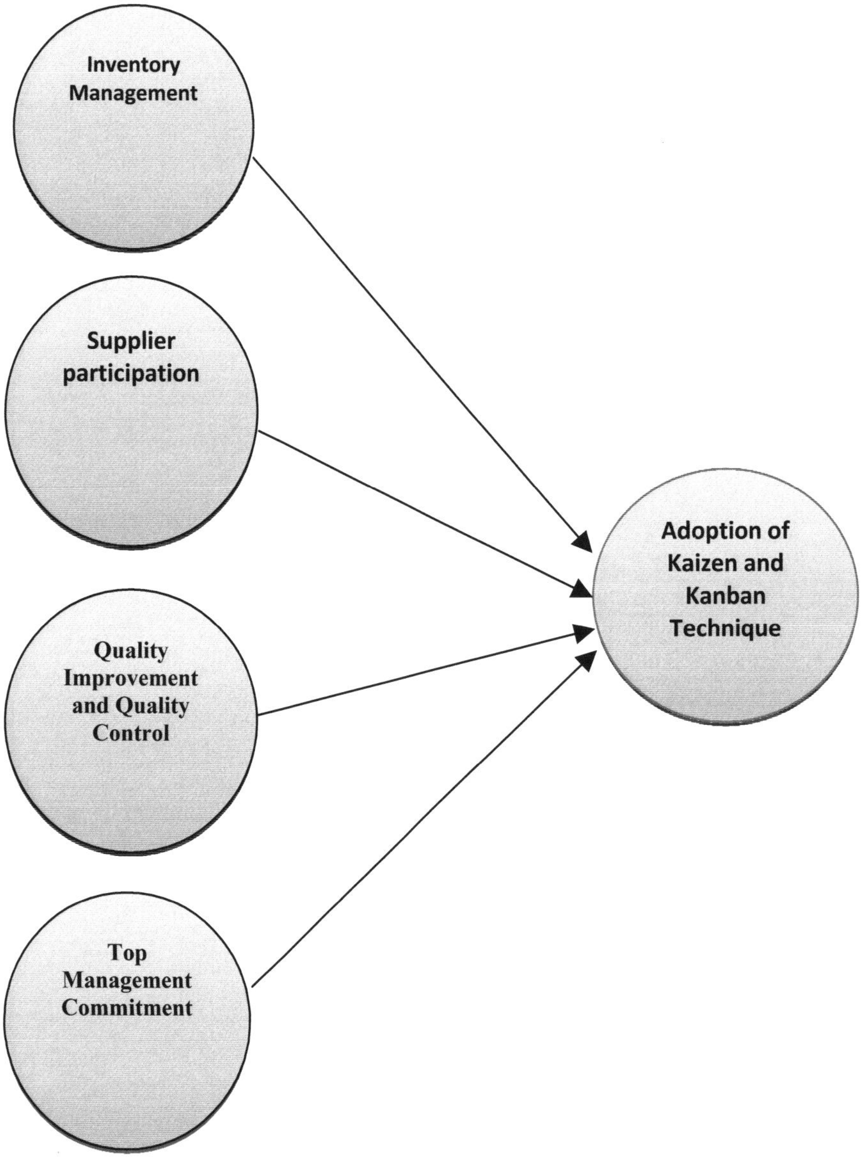

Figure 1 Theoretical model

RESEARCH QUESTIONS

Major research question that pursued through this research was:

1. Does any direct and positive relationship exist between Inventory, Supplier Participation, Quality Improvement and Quality Control And Employee Participation and Top Management Application or Adoption of Kaizen and Kanban Technique

RESEARCH OBJECTIVES

The main objectives that pursued through this research are as follows:

1. To examine the relationship between Inventory and Application or Adoption of Kaizen and Kanban Technique.
2. To analyze the relationship between Supplier Participation and Application or Adoption of Kaizen and Kanban Technique.
3. To ascertain the relationship between Quality Improvement and Quality Control and Application or Adoption of Kaizen and Kanban Technique.
4. To evaluate the relationship between Employee Participation and Top Management and Application or Adoption of Kaizen and Kanban Technique.

RESEARCH HYPOTHESIS

The main hypotheses that pursued through this research are as follows:

1. Relationship exists among Inventory and Application or Adoption of Kaizen and Kanban Technique.
2. Relationship exists between Supplier Participation and Application or Adoption of Kaizen and Kanban Technique.
3. Relationship exists among Quality Improvement and Quality Control and Application or Adoption of Kaizen and Kanban Technique.
4. Relationship exists between Employee Participation and Top Management and Application or Adoption of Kaizen and Kanban Technique.

MEASUREMENTS OF CONSTRUCTS

Most of the measurements that have been utilized in this current study were adapted from past-established instruments contextualizing the topic to Omani industrial establishments. The instruments further contextualized to Kaizen and Kanban in Sultanate of Oman. All measurement items (38 items) coming from each construct and its Cronbach alpha level0.60 passes the minimum requirement (Nunnally, 1960).

Table 1 Measurement of Constructs

No	Instruments	No of Items	Scale (Likert)	Reliability	Authors
1	Inventory Management	7	5 point scale	0.90	Gunasekaran and Patel (2001)
2	Supplier participation	21	5 point scale	0.89	Vazquez et.al (2004), Walter. Mentzer and Croxton (2002), and Berry and Parasuraman (1991)
3	Quality Improvement and Quality Control	5	5 point scale	0.87	Yusof (1999).

4	Top Management Commitment	5	7 point scale	0.86	Saraph, ; Benson, ; Schroeder (1989)

Table 2 Factor Analysis

	Code	1(I)	2(SP)	3(QIQC)	4(TMC)
Inventory	I 1	0.71			
	I 2	0.60			
	I 3	0.66			
	I 4	0.62			
	I 5	0.70			
	I 6	0.71			
	I 7	0.70			
	SP 1		0.71		
Supplier Participation	SP 2		0.70		
	SP 3		0.66		
	SP 4		0.62		
	SP 5		0.75		
	SP 6		0.63		
	SP 7		0.65		
	SP 8		0.65		
	SP 9		0.70		
	SP 10		0.66		
	SP 11		0.63		
	SP12		0.61		
	SP 13		0.71		
	SP 14		0.59		
	SP 15		0.58		
	SP 16		0.66		
	SP 17		0.65		
	SP 18		0.64		
	SP 19		0.63		
	SP 20		0.61		
	SP 21		0.62		
Quality Improvement and Quality Control	QIQC 1			0.66	
	QIQC 2			0.69	
	QIQC 3			0.55	
	QIQC 4			0.58	
	QIQC 5			0.60	
	TMC 1				0.66
Employee Participation and Top Management Commitment	TMC 2				0.64
	TMC 3				0.60
	TMC 4				0.61
	TMC 5				0.65

Variance
Kaiser-Meyer-Olkin Measure of Sampling Adequacy = 0.920
Approximate. Chi-Square = 1311.2
DF = 46 (boot strapping 500)
Sig. = 0.000
Variance = 70.291

Table 3 Correlations service quality

	Correlations			
	IM	SP	QI&QC	TMC
IM	1			
SP	.599**	1		
QI & QC	.623**	.612**	1	
TMC	.609**	.604**	.580**	1

**.Correlation is significant at the 0.01 level (2-tailed)

The result in Table 3 on inventory management indicates significant and positive relation to adoption of Kaizen and Kanban (r=0.599, P< 0.01). The table further indicate similar reationship between supplier participation and adoption of Kaizen and Kanban (r=0.623, P< 0.01) and quality improvement and quality control with the adoption of Kaizen and Kanban Further the result between top management support and adoption of Kaizen and Kanban also indicates significant and positive relation (r=-0.609, P< 0.01).

Internal consistency

Internal consistency was used to ascertain the measurement model through the internal consistency. It also ascertain whether the set of indicators are reliable or not relaible. To overcome some of the limitations of using Cronbach's alpha, the composite reliability was suggested in the Structural Equation Modeling (SEM) literature (Anderson & Gerbing, 1988). For safety purpose, this particular study utilized both Cronbach's alpha and the composite reliability measures in order to have more accurate findings.

Table 4 Measurement Model Quality Criteria

	Cronbach Alpha	Composite Reliability	AVE	Communality
I	0.91	0.92	0.57	0.54
SP	0.84	0.93	0.61	0.58
QIQC	0.81	0.91	0.64	0.61
EPTMC	0.89	0.93	0.59	0.58

Table 4 demonstrates the results indicating composite reliability and the values of cronbach alpha for all the relective constructs. The results shown in Table 1 indicate that all study construct meet the statisfactory level of composite reliability and cronbach alpha meeting the minimum cut off value 0.7.

METHODOLOGY

Research Design

This empirical study focuses on application of Kanban and Kaizen techniques of the business houses of Rusayl Industrial Estate. The nature of study is fact finding and research based. It follows cross sectional descriptive study design as its plan of action. Since the research on Kanban and Kaizen techniques followed positivist philosophy and the phenomenon under study is objective, an epistemology which is quantitative research is more suitable to explain the relation between independent and dependent variable selected under the study.

Population

Out of the eight Industrial Estates, the study chosen Rusayl Industrial Estate as the area of study since it is the oldest one and has 230 business plots inside the IE (Oman Directory 2016 published by PEIE). This particular research is aimed at investigating the intention to adopt Kanban and Kaizen techniques among industries.

Sampling

The universe of the study is 230 business plots of Rusayl Industrial Estate. Out of this infrastructure providing facility is taken care of by three banks, one water supply company and an electricity provider. Sampling concerns the determination of the study's population, sample frame, and sample size. Each subject will be discussed in the ensuing paragraphs. This study employed a purposive sampling technique to collect 46 functional survey forms from the staffs who work at in the business plots of Rusayl Industrial Estate.

ANALYSIS AND RESULTS

Descriptive Statistics Result

Table 5 Result of the Descriptive Statistical Analysis

Descriptive Statistics					
	N	**Minimum**	**Maximum**	**Mean**	**SD**
IM	46	1	5	4.33	1.004
SP	46	1	5	4.12	1.032
QI & QC	46	1	5	4.03	1.011
TMC	46	1	7	6.29	1.040
Valid N (list wise)	46				

The Table 5 provides values of the means and the standard deviation values for the latent variables, construct wise with the minimum and maximun scores. The mean of the independent variables following into ranging value between 6.03-3.99 and the rest of the variables also shown incline towards agreeable level. The standard deviation values show that there are not much deviation from mean in the data.

Table 6 Details of Business Houses classification according business type.

Sl.No	Type of Industry (In detail)	Population (230 units)
1	Engineering & Electrical Industry	*26*
2	Construction, Building, Fire safety equipment.	*7*
3	Printing, Plating, Metal sheets, Glass and Mattress Manufacturing	*8*
4	Industrial Products and Packaging Industry	*11*
5	Trading and contracting	*20*
6	Marble & Cement, Furniture, Tiles	*17*
7	Plastic & Cables	*11*
8	Food & Water, Food processing	*17*
9	Glass, Pipes & Fibers	*5*
10	Bank Investments, trading technology	*13*
11	Chemicals Industry & Pharmacy	*10*
12	Oilfield technology, Bitumen	*15*
13	Travel, Tourism Logistics	*7*
14	Warehouse, Water treatment company	*6*
15	Aluminum iron and steel fabrication	*16*
16	Shoe Garments and leather	*9*
17	Others	*23*
18	Telecommunication, health centre, royal court, training centre, electricity together	*9*
	TOTAL	*230*

Table 7 Details of Business Houses: Classification according Capital Type.

Sl.No	Capital Contributed by:	Population (230 units)
1	Government	*Nil*
2	Private	*227*
3	Shared	*3*
	TOTAL	*230*

Table 8 Details of Business Houses: Classification according Legal Status.

Sl.No	Legal Status	Population (230 units)
1	SAOG	*16*
2	LLC	*157*
3	Limited Partnership	*2*
4	General Partnership	*1*
5	Sole Proprietorship	*8*
6	SAOC	*2*
7	Branch Office	*0*
8	Representative Office	*0*
9	Government Office	*0*
10	Others	*44*
	TOTAL	*230*

Table 9 Details of Business Houses: Classification according Business License.

Sl.No	Industrial License	Population (230 units)
1	Industrial	*164*
2	Commercial	*24*
3	Residential	*1*
4	Service	*41*
5	Agricultural	*0*
6	Animal Husbandry	*0*
7	Fishy	*0*
8	Government	*0*
	TOTAL	*230*

(Source: https://ems.peie.om/newsDir/directory.aspx?lang=en.US)

For our study we have concentrated on the classification according to business type since that classification helps us to identify and synthesize data collected regarding application of Kaizen and Kanban techniques. The determinants of adopting and applying Kaizen or Kanban techniques has been listed out after discussion with 144 respondents who belonged to various positions in the business houses of The Rusayl Industrial Estate. It is ranked based on Garret's Ranking Table. Various determinants of Kaizen and Kanban Technique have been ranked based on the intensity and analyzed using Garrett's ranking table. The calculations are as under. This shows the ranks given for the determinants in the order of relevance to adopt the Japanese technique.

Table 10 Determinants of application of Kaizen and Kanban Technique – Scores

S.No	Determinant	1	2	3	4	5	6	7	8	Total
1	Inventory Management Problems	65	22	26	17	8	3	2	1	144
2	Quality control related issues	12	68	57	7	0	0	0	0	144
3	Inflationary tendency	16	31	37	59	0	0	0	1	144
4	High price fluctuations	15	4	3	31	28	27	20	16	144
5	Warehousing facility cost	13	2	2	3	51	34	29	10	144
6	Supplier Participation	8	11	6	14	27	29	24	25	144
7	Employee Participation	6	2	1	6	14	29	50	36	144
8	Expat workforce problems	9	4	12	7	14	23	20	55	144
	Garrett's Table Value	79	67	59	53	46	40	32	21	

Source: Primary data

Table 11 Garret's Ranking Table: Determinants of Application of Kaizen / Kanban Technique

S.No	Reason	Total Score	Average	Rank
1	Inventory Management Problems	9617	66.79	I
2	Quality control related issues	9238	64.15	II
3	Inflationary tendency	8672	60.22	III
4	High price fluctuations	6617	45.95	IV
5	Warehousing facility cost	6282	43.63	V
6	Supplier Participation	6160	42.78	VI

7	Employee Participation	5145	35.73	VIII
8	Expat workforce problems	5417	37.62	VIII

It is observed from the Table that "Inventory Management Problems" has been ranked as the first determinant of adoption of kaizen and Kanban technique followed by quality control and inflationary tendency issues.

STRUCTURAL MODEL

Table 12 Structural Model

Hypothetical Relationship	Path Coefficient	Absolute t statistic value	Value of R^2	Value of Q^2
IM → IAK&K	0.583	19.221	0.680	0.599
SP → IAK&K	0.600	17.990	0.613	0.553
QI&QC → IAK&K	0.615	20.113	0.699	0.601
TMC → IAK&K	0.583	16.539	0. 590	0.499
Over all			0.698	0.590

In this section, the result of the structural model and tests of hypotheses for this study will be presented. Following Hair et al. (2011; 2012), this section is concerned using the standard bootstrapping process with a number of five thousand bootstrap sample on three seventy eight cases to assess the importance of the path coefficients. Figure 4.9 illustrates the estimation for the full structural model. To understand the relationship between the constructs, SEM-PLS structural model analysis was conducted. The contribution of each construct of the exogenous variable is represented by the standardized beta values within the PLS structural model (Chin, 1998).

OVER ALL MODEL

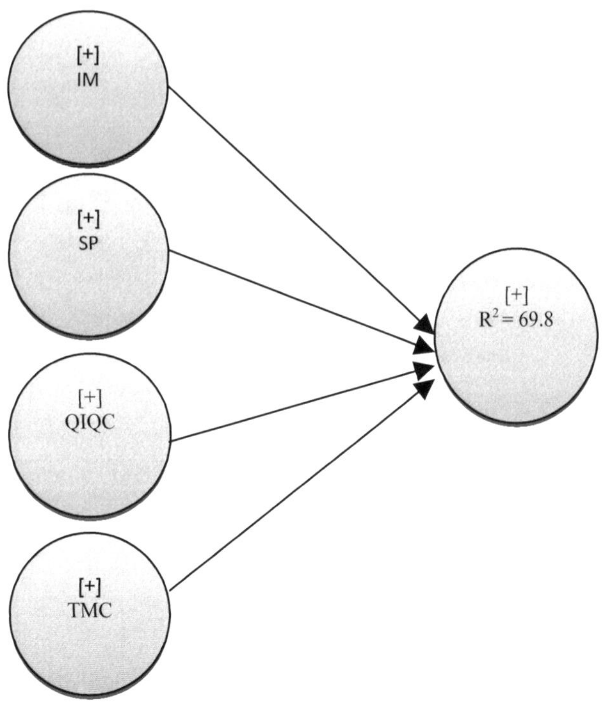

Figure 2 Structural model

Table 13 Summary of Hypothesis Testing

Research Questions (RQ) & Hypothesis	Assessment
RQ1: Does any relationship exist between inventory, supplier participation, quality improvement and quality control and Employee Participation and Top Management application or adoption of Kaizen and Kanban Technique.	
H1: Relationship exists among supplier participation and application or adoption of Kaizen and Kanban Technique. H1: IM→IAK&K	Supported
H2:Relationship exists between supplier participation and application or adoption of Kaizen and Kanban Technique. H2: IM→IAK&K	Supported
H3: Relationship exists among quality improvement and quality control and application or adoption of Kaizen H3: IM→IAK&K	Supported
H4: Relationship exists between Employee Participation and Top Management and application or adoption of Kaizen and Kanban Technique. H4: IM→IAK&K	Supported

It is necessary to establish the relations between independent and dependent variables, in order to analyses the cause effect relation. The study followed Structural Equation Modeling (SEM) to analyze the cause effect relation between the independent and dependent variables selected for the study (Chin, 1996; Haenlein & Kaplan, 2004; Statsoft, 2013). Significant condition for assessing the structural model in PLS-SEM is the R squared value, which is also known as the coefficient of determination (Hair *et al.,* 2011; Hair *et al.,* 2012; Henseler & Fassott, 2010). Chin (1998) suggests that the R-squared values of 0.67, 0.33, and 0.19 in PLS-SEM can be considered as substantial, moderate, and weak, respectively. In particular, a cross-validated redundancy measure (Q^2) was applied to assess the predictive relevance of the research model (Chin, 2010; Geisser, 1974; Hair *et al.,* 2013; Ringle, Sarstedt, & Straub, 2012; Stone, 1974). According to Henseler & Fassott (2010), a research model with Q2 statistic (s) greater than zero is considered to have predictive relevance. Additionally, a research model with higher positive Q2 values suggests more predictive relevance. The table 12 clearly indicates the extrapolative significance of the model generated in this research. This particular study has posed 4 hypotheses involving the direct and indirect analysis. The direct analysis was related to 4 hypotheses, based on one research question. The composite model which has come out the overall analysis indicates that the entire hypothesis considered for the study is well established the causation (69.8%).

DISCUSSION

This discussion is on the result of research in 4 objectives framed for the study. The purpose of this research was to get better insight into the industrial executive's attitude towards Kaizen and Kanban adoption and acceptance. Theoretical outcome of the study provides observed evidence for the direct relationship between Inventory management, supplier participation, quality improvement and quality control and top management support in the relationship with Kaizen and Kanban adoption and acceptance in the Rusayl Industrial Estate. The better the Inventory management, supplier participation, quality improvement and quality control and top management support, in Kanban adoption and acceptance, higher will be their intention to adopt it. Several past studies have verified the support of this relationship. There are several supplies used in operations as well as finished goods in the form of raw materials; work in process; etc. (Muller 2011, 1). The objective of inventory management is betting best of requirements which satisfy the consumers. The management has to minimize the operation cost by extending best of requirements with minimal cost. Kanban tries to keep the inventory cost at lowest extend with the support of right equipments, facilitating right time period before productions starts. There is effective control from the point of assembly to the manufacturing operations that precede it. (DuBrin, 2008).

The current study clearly indicates the connection between inventory management and the need of adoption of Kaizen or Kanban for minimizing total operational cost. The Kanban process where inventories are narrowed by providing these when needed in production. Effective supplier participation can bring down the cost of operation in business. The present study clearly indicates the relationship between inventory management and the need of adoption of Kaizen or Kanban for minimizing total working cost. Companies which are keen in quality control and quality improvement make use of Kanban or Kaizen tools to ensure the elimination of the waste as well as reducing the operating cost. The present study clearly indicates the relationship between quality improvement and quality control effort and the need of adoption of Kaizen or Kanban for minimizing total working cost. (Surendra et al., 1999).The 4th hypothesis considered for this research was to establish the relationship between top management support and adoption of Kaizen or Kanban tools. The study clearly establishes the fact that the top management support is a major factor which will lead to a decision whether to adoption of Kaizen or Kanban tools. It appears from the result that issues related to organizational leader's support is primarily seen either as hurdle or as smooth way in implementing Kaizen or Kanban tools.

Oman as a national objective is encouraging PEIE. Rusayl Industrial Estate is well developed with respect to its infrastructure. Our study on the Rusayl Industrial Estate with respect to application and adoption of Kaizen and Kanban Techniques by the business houses, and to state the determinants of the technique brought forth certain findings. Huda (Huda, (1992)) found that Kaizen has no cultural dependence and this finding is matching with our ranking the determinants of Kaizen and Kanban adoption and acceptance where the cultural aspects were not cited at all. Rahmana, (2013), found that Kanban system is one of the strategies of lean production which is in line with the findings of the present study that inventory influences the adoption or application of Kaizen and Kanban techniques. The present study analyses and ranks the determinants of adoption or application of Kanban and Kaizen techniques and found that quality control influences the adoption of this technique and this is also in line with the findings of (Lam, 2006) where they found that Lean manufacturing program designed to increase product quality. But our study defies the findings of the study carried out by (Singh, 2009) where they found that Kaizen strives to empower the workers whilst our study brings forth that supplier or employee participation is not influenced by the application or adoption of Kaizen or Kanban techniques. From this though we don't infer that suppliers or employees can influence Kaizen but Kaizen technique can empower the employees. Thus our study ranks inventory management and quality control as factors which influence the adoption and application of Kanban or Kaizen technique.

IMPLICATIONS

The findings of the study and the discussion which followed clearly indicate inventory management problems and quality control issues influences the adoption and application of Kaizen and Kanban techniques in manufacturing and service rendering. The ranking of determinants denotes that inventory management, quality control and inflationary tendencies all lead to adoption of Kaizen and Kanban techniques. If any orientation is given to the production departments by the Management Accountants of Professional Institutes the application of Kaizen and Kanban techniques would help them to work more efficiently and conduct their business without much wastage.

SUGGESTIONS

The study suggests following measures to the concerned authorities to attract more organizations to apply Kaizen and Kanban techniques are as follows:

1. The production process needs to be designed in a way to suit Kaizen and Kanban techniques.
2. For new enterprises which are on the verge of beginning the Kaizen costing would be of immense help and so should be adopted.
3. There is a need to extend more training on the lean manufacturing and continuous improvement need to be given to the entrepreneurs.
4. A centralized organization to be established to facilitate Kaizen manufacturing method of the products of the organizations.
5. Provide technical and technological know-how to the new entrepreneurs in the RIE, and support them with all kinds of information that they might need.
6. It is envisaged that the financial institutions should extend better helping hand to the entrepreneurs by giving low interest rates for based on feasibility study.
7. Making use of Kaizen and Kanban techniques difficult process of procuring g raw materials to be avoided.
8. Knowledge sharing is a must. At institutional or industrial level proper Knowledge sharing from experts to be encouraged.
9. Industrial festivals and other social gathering events will support the budding industrial entrepreneurs to motivate them towards Kaizen and Kanban techniques adoption.

CONCLUSION

The purpose of the research was to analyze the link among inventory management, supplier participation, quality improvement and quality control and top management participation on adoption of Kaizen and Kanban tools by the business houses in the industrial establishment of Rusayl Industrial Estate of Sultanate of Oman. Though there was lot of studies conducted on the Industrial Estate our study is unique since none of the researchers attempted to look into the adoption of these techniques in the present scenario. The study followed one research question to operate the research. The findings clearly indicate that there is significant relationship between inventory management, supplier participation, quality improvement and quality control and top management participation on adoption of Kaizen and Kanban tools. The findings of the present study and the suggestions extended can be further boosted through massive campaigns and orientation programs addressing the benefits of these techniques in manufacturing and service delivering. Professional Management Accountants are the best sources of knowledge as far as the application and implementation is concerned. The study paves better insight for the top management of industrial units establishes the cause effect relationship by sighting the implications.

REFERENCE

Anderson, J. C., & Gerbing, D.W. (1988). Structural Equation Modeling in Practice: A Review and Recommended Two-Step Approach. *Psychological Bulletin, 103*(3), 411-423.

Berry, L.L., & Parasuraman, A. (1991). *Marketing Services: Competing through Quality*, Free Press, New York, NY.

Burch, M.K. (2008). Lean Longevity: Kaizen Events and Determinants of Sustainable Improvement. In Burch, M. K. *leans longevity: kaizen events and determinants of sustainable improvement.* ProQuest LLC.

Chin, W.W. (2010). How to write up and report PLS analyses. In V.E. Vinzi, W. W. Chin, J. Henseler, & H. Wang (Eds.), *Handbook of partial least squares: Concepts, methods and applications in marketing and related fields* (pp. 655-690).Berlin: Springer.

Chin, W. W., Marcolin, B. L., & Newsted, P.R. (1996). A partial least squares latent variable modelling approach for measuring interaction effects: Results from a Monte Carlo simulation study and voice mail emotion/adoption study. Paper presented at the 17th *International Conference on Information Systems*, Cleveland, OH.

Vaus, D. (2002). *Surveys in Social Research* (5th ed.). London: Routledge.

Dubrin, A. J. (2008). *Essentials of Management*. Cengage Learning.

Geisser, S. (1974). A Predictive Approach to the Random Effects Model, *Biometrika 61*(1), 101-107.

Gunasekaran, A. Patel, C. & McGaughey, R. (2004). A framework for supply chain performance measurement, *International Journal of Production Economics, 87*(3), pp.

Haenlein, Michael & Kaplan, A. (2004). A beginner's guide to partial least squares analysis, *Understanding Statistics, 3*(4), 283 – 97

Hair, J., Ringle, C. & Sarstedt, M. (2011). PLS-SEM: Indeed a silver bullet. *Journal of Marketing Theory and Practice, 19*(2), 139-151.

Hair, J.F., Hult, G.T.M., Ringle, C.M., & Sarstedt, M. (2013). A *Primer on Partial Least Squares Structural Equation Modeling*. Sage, Thousand Oaks.

Hair, J.F., Sarstedt, M., Ringle, C.M. & Mena, J.E.(2012). An Assessment of the Use of Partial Least Squares Structural Equation Modeling in Marketing Research. *Journal of the Academy of Marketing Science 40*(3), 414-433.

Henseler, J. & Fassott, G. (2010). Testing moderating effects in PLS path models: an illustration of available procedures. In: Esposito Vinzi, Vincenzo; Chin, Wynne W.; Henseler, Jörg; Wang, Huiwen (Eds.). *Handbook of Partial Least Squares: Concepts, Methods and Applications* (Springer Handbooks of Computational Statistics Series, vol. II). Heidelberg, Dordrecht, London, New York: Springer, pp. 713-735, doi:10.1007/978-3-540-32827-8_31

Huda, F. (1992). Kaizen: the applicability of Japanese Techniques in IT. *Software Quality Journal,* (1), 9-26. doi:https://doi.org/10.1007/BF01720166

Jayeola1, S.A. (2012). Kaizen Cost Management Technique and Profitability of Small and Medium Scale Enterprises (SMEs) in Ogun State, Nigeria.*Research Journal of Finance and Accounting*, 3(5). 03-113. ISSN 2222-1697 (Paper) ISSN 2222-2847 (Online).

Lam, T.A. (2006). Managerial Accounting and Continuous Improvement Initiatives: A Retrospective and Framework. *Journal of Managerial Issues, 18*(2), 157-174

Müller, M. (2011). *Essentials of Inventory Management.* AMACOM Diverse American Management Association.

Nunnally, J. (1960). The place of statistics in psychology. *Educational and Psychological Measurement, 20*, 641-650.

Rahmana, S. M. (2013). Lean Manufacturing Case Study with Kanban System Implementation. *International Conference on Economics and Business Research* 2013 (ICEBR 2013), 7, pp. 174 – 180. (retrieved from www.sciencedirect.com)

Ringle, C.M., Sarstedt, M., & Straub, D.W. 2012. A Critical Look at the Use PLS-SEM in *MIS Quarterly 36*(1), 3-24.

Saraph, J.V., Benson, P.G., & Schroeder, R.G. (1989). An Instrument for Measuring the Critical Factors of Quality Management. *Decision Sciences*, *20*(4), 810-829. DOI: 10.1111/j.1540-5915.1989.tb01421.x

Singh, J.S. (2009, May). Kaizen Philosophy: A Review of Literature. IUP Journal of Operations Management, 51-72. Retrieved from https://search.proquest.com/openview/8cfdcced 5aa335d1774 65b89af8bc359/1?pq-origsite=gscholar&cbl=54466

Stone, M. (1974). Cross validatory choice and assessment of statistical predictions, *Journal of the Royal Statistical Society*, *36*(2), 111-147.

Surendra, M.G.,Yousef, A.Y., & Ronal, F.P. (1999). Flexible Kanban system. International *Journal of Operations and Production Management. 19*(10), 1065-1093.

Vazquez, R., Iglesias, V. & Alvarez-Gonzalez, L. I. (2005). Distribution Channel relationships: The Conditions and Strategic Outcomes of Cooperation between manufacturer and distributor. *Journal International Review of Retail, Distribution & Consumer Research*, 15,125-150

Walter., Mentzer., & Croxton (2002). Cited in: *Information sharing, inventory management and customer satisfaction in the downstream chain of manufacturing firms in Uganda,*: Research report submitted to Makerere university business school in partial fulfillment of the requirements of a master of science in procurement and supplies management of Makerere university. October, 2010

Yusof, S.M. & Aspinwall, E.M. (1999) Critical Success factors for total quality management implementation in small and medium enterprises, *Total Quality Management, 10*, 803-809

From E-learning and M-learning to Virtual and Augmented Reality : A Comparative Study

Rula Al-Azawi, IEEE member
Gulf College, Oman
rula@gulfcollege.edu.om
rulaalazzawi@yahoo.com

Indu Pillai
Gulf College, Oman
indu@gulfcollege.edu.om

Raziyeh Moghaddas
Gulf College, Oman
Raziyeh@gulfcollege.edu.om

Ronald S. Cordova
Gulf College, Oman
ronald@gulfcollege.edu.om

Abstract—**The popularity of the Internet as an information and education resource has grown extremely in this century. This paper discuss the new trend of technology in education which is based on delivered materials which is not based on the availability of teacher or lecturer.**
Educational technologies currently focus on dealing with electronic learning (E-Learning) and mobile learning (M-Learning) as a useful educational tool. Many studies start to give big attention on the next step in Virtual and Augmented Reality as useful tools for education, training, simulation.. etc.
This research paper presents a comparative study dealing with variety technologies in education (e-learning, m-learning, Virtual Reality, Augmented Reality). Furthermore, advantages and disadvantages for each educational technology has been addressed. The main goal for this paper is to assist teachers to be motivated appreciate by adding new technologies in their teaching method by selecting a suitable technology based on teaching requirements and materials.

***Keywords*—E-learning , mobile learning , M-learning, Virtual Reality , Augmented Reality**

I. Introduction

Most of the current researchers and lecturers focused on exploring innovative technique in order to deliver their material online, motivate students remotely, get more engagement with the student and enhance the E-learning experience.
In higher education academy, most of the academic lecturers are interested in E-learning and how they will be able to deliver their course materials online. Of course, many universities are using blackboard and moodles as a communication and educational online technology.
More specifically, in this paper first we are interested to present a comparison between different on-line materials delivered in interactive and simulated way to students. Secondly, our aim is to create an on-line community similar to an actual university setup for on-line students.
In the context of distance education which provides opportunities and advantages in teaching, learning, delivering and facilitation of teaching method by teacher, trainer, instructor or lecturer. The learners like students increases their collaboration and are more interested in the new way than traditional teaching and learning methods.
Presently, the need for usage of technologies which removes the boundaries of time and location that increases day by day when information and access to information gains importance. Effect of electronic learning to education is an issue to be researched in order to provide lifelong learning[1].
In this paper, we will explain in details about different distance learning terms such as E-learning, M-learning, Virtual Reality and Augmented Reality and presents the usage of each of them and also the strengths and weaknesses in a comparative study. The rest of the paper is organized and structured as follows: section two, presents an explanation of different educational technologies such as E-learning, M-learning and also the new trends of educational technologies such as Virtual and Augmented Reality. Section three, discusses the difference between them and shows the advantages and disadvantages for each educational technology. Our conclusion and future work is presented in section four.

II. Educational Technology

The traditional educational methodology is made in actual classroom where teacher is present and delivers materials to a group of students. While the educational technology depends mainly of teacher and the students must physically participate in the learning process. Regardless of obvious advantages as a direct contact between a teacher and students and immediate feedback the traditional classroom education has many disadvantages. For example, if the student has no ability to take part in some lesson he or she will miss the training materials. These disadvantages lead us to search for new and more effective educational methods [2].
In this century, technology influences the transformation of the teaching and learning approach among educators and learners. Meanwhile, technology also decreases the barriers or restriction in education, and promote freedom in the teaching and learning process [4] [3].
The fast growth on technology and communication technologies creates new trends in educational forms. Initially, the educational technology focuses on the D-learning. In D-learning the personal contact between a teacher and a student is replaced by electronic mails or any type of communication based on modern technology. Here the students used to communicate with the teacher at their own pace. As the development of Internet continued popularity in the late 80s and early 90s, E-learning, overtook the choice as a remote education tool. In E-learning the students have a learning experience using the multimedia such as text, audio, video with personal interaction with their teachers and peers. Currently M-learning is mostly preferred technology which is distance learning using portable wireless devices with constant Internet access.
The next generation of educational technology will be based on Virtual and Augmented Reality. Figure1 illustrates the relation between different educational technologies. In the next section, we will explain each terms related to educational technology

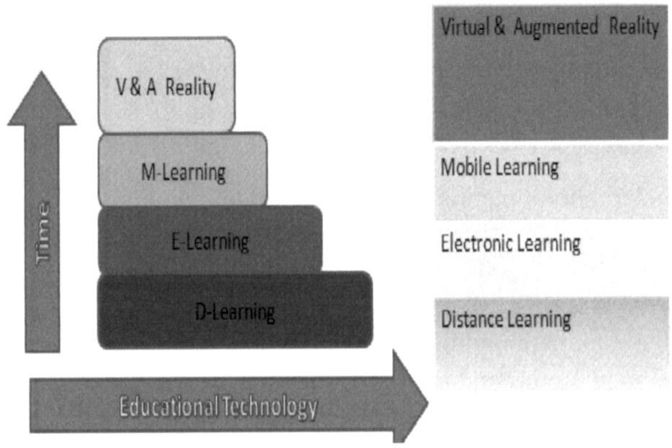

Fig. 1: The usage of different Educational Technology terms over the Time

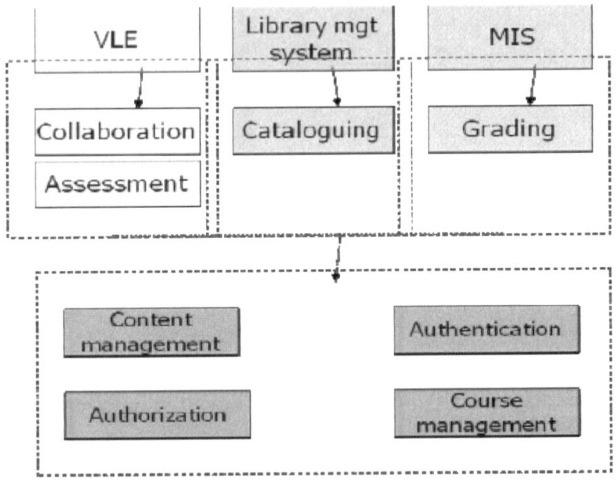

Fig. 2: Architecture of a simplified Learning System

starting from E-learning , M-learning reaching to Virtual and Augmented Reality. Moreover, an explanation on how of those technology ables to be used as part of teaching, training, learning and simulation.

A. *E-learning*

In the past, the term E-learning has been used for any method of learning that is based on electronic delivery method. Currently, E-learning usually refers to on-line courses.
The E-learning is a new method for Virtual learning based on computer and new technologies. It includes on-line learning, on-line based training, open universities and classrooms, digital collaboration and technology assisted distance learning. The progress and support for the E-learning environment will extremely change how we use learning technologies. This learning style will provide more capability and more freedom for the learners. They would indicate or suggest redrawing the balance between traditional learning and learning from worldwide .To apply change is very difficult, but probably the immediate development and execution of new technologies will bring a dramatic change in our educational field [5].
E-learning is an Internet-based learning process, using Internet technology to design, implement, select, manage, support and extend learning, which will not replace traditional education methods, but will greatly improve the efficiency of education. As e-learning has a lot of advantages like flexibility, diversity, measurement, opening and so on, it will become a primary way for learning in the new century as in Figure 2 [6].

B. *M-learning*

M-learning takes in hand the aspect of "mobility" which is capable of being moved or moveable. As a consequence, the learning can take place anytime, anywhere, without any time and accessibility limitations of the E-learning technology. Furthermore, M-learning opens learning windows on a large scale portable devices which provide mobility and access to information "on the go," such as smart phones, tablets, PDAs, etc. [7].

M-learning is an evolved advancement of D-learning and E-learning. Naturally M-learning is a sort of existing D-learning and E-learning. Mlearning also assures equal educational opportunity for disabled people[2].
M-learning enhances the learning process and becomes more interesting through the ability of connectivity with mobile devices, such as mobile Internet, Blue-tooth and Wi-Fi (local area wireless computer networking)[8]. These functions enable students to conduct mobile learning without the limitation of location and time [9]. Besides that, M- learning also allows students to conduct group discussions in Cloud and social network applications, without meeting each other and it does not limit the number of participants [9]. Thus, the teachers and students are able to provide instant response when the discussion or lesson is ongoing.
On the other hand, the teachers can share information such as picture or video in the group to enhance the student's social work. M-learning is advancement in terms of technology in comparison to E-earning and Distance Education/learning. Research in Electronic and Mobile Learning focuses on: Technical Feasibility Analysis, System Terminal Software Development, Curriculum Development, Teaching/Learning Modes, Standard of Learning System and Theoretical Study[10] [11]. M-learning emerges with person-to-person communication done via mobile devices [7]. Mobile technology is the foundational infrastructure to sustain M-learning. The portable and autonomous devices have made some researchers defined mobile learning based on the physical dimensions of the devices. Also M-learning is defined with more focus on the mobility of the learners [12]. According to Kukulska-Hulme [12], M-learning can be formal or informal and learner independent to choose what and when to learn. Considering these facts and ideas, the main characteristics of M-learning are the learners mobility and learning virtually anywhere and anytime using mobile devices. How to teach and learn through M-learning became researchers interest as well [13] [14].
If we analyze mobile education in terms of its advantages, we can range them as follows;

1) Life-long learning,

2) Learning inadvertently

3) Learning in the time of need,

4) Learning independent of time and location,

5) Learning adjusted according to location and circumstances [15]

C. Virtual Reality (VR)

Virtual Reality is used to break the boundaries of formal education. Virtual Reality has been used in education sector in order to improve the on-line learning. Some aspects give students the sense of presence in an immersive world, and bringing a new way to interact with colleagues and teachers when they are not in the same place [16].

The main objective of Virtual Reality is to achieve the concept of immersion. Jennett [17] defines immersion as the involvement with the Virtual condition which causes the absence of consciousness of the real time and real world, causing an affair of "being" in this Virtual universe. Jennet worked on this definition considering game context, yet the embodiment of such definition could be investigated while examining any Virtual Reality environment. "Immersion" in Virtual Reality is usually utilized as a part of regard of "spatial immersion", the impression of quality in a nonphysical world, which is caused by the stimulation of images, animations, sounds and others interactivity aspects. The user can investigate the Virtual environment in 360 degrees and can cooperate with things inside this Virtual world; the user sees and hears this Virtual environment as a genuine place, and it is the thing that we call the "feeling of essence". One may contend that a full and finish "immersion" may include every human sense, while the current Virtual technology concentrates chiefly in sight and hearing. This is valid; the present business Virtual Reality technology does not include other human detects, for example, taste, smell, and touch.

In the past decades, the early stage of Virtual Reality has been integrated development based on the interactive 3D graphics, user interface, and visual simulation [18]. Virtual Reality attracts people's attention, and the technology has been applied in medicine, industry, education, video games, tourism and other industries [19].

Many researchers and examples of using Virtual Reality has been presented such as Du [20], how user Virtual Reality to increase motivation of study as a part of learning process while Griol [21] uses Virtual Reality to improve learning computer science by using open simulator. Furthermore, Mathur[22] offers a low cost demo for medical trainer and instructional purposes.

Those experiments and researches show significant evidences that Virtual Reality has characteristics of an immersive technology that could provide a more active and engaging learning environment to motivate students and keep their attention[16].

D. Augmented Reality

Augmented Reality is a technology that connects the computer world to the human world [23]. While other authors define it as a technology that allows users to see the real world with computer generated objects superimposed[24], [25]. According to Mazuryk [26], initially, the Augmented Reality was used for military purpose to develop an advanced flight simulator by the US Air Force Armstrong Medical Research Laboratories in 1982. Nowadays, Augmented Reality is also implemented in the education field [27][28].

The term Augmented Reality presented in 1950 by Morton Heilig who created a Virtual Reality system which had the relevant elements such as an environment but without interaction [26]. The Augmented Reality could enhance the teaching and learning experience by engaging the learner to explore the world by using multimedia elements to support the investigation of the surrounding.

The students believed that the integration of technologies would help them in their learning process. Therefore, educators have begun to seek technologies that have the potential to be integrated in education in order to help students learn actively and to improve their understanding especially in Science subjects [27]. Augmented Reality is a new way to improve the 3D shape instead of using wood object. There are many advantages of using Augmented Reality in education mentions by many authors as shown below:

1) Supports seamless interaction between real and virtual environments and allows the use of a tangible interface metaphor for object manipulation[29].

2) Provide instructors with a way to strengthen students understanding in the classroom by Augmenting physical props with Virtual annotations and illustrations [30].

3) Creates a learning experience that is linked to the formal classroom, so that students can learn outside of class hours and outside of school limits [31].

4) Enables the visualization of interactions among amino acids and protein building processes as static 2D/3D images and 3D dynamic images (animations)[32].

III. Comparative Study and Critical Analysis

Four types of educational technology were discussed in the previous section of this paper. There are some differences and similarity between aforementioned technologies in different criteria such as equipment, place, mobility, interaction, connectivity and display. Our comparative study will be divided into two categories, firstly, we will compare and analysis the advantage and usage of E-learning and M-learning. Secondly, we will compare between both Augmented Reality and Virtual Reality with M-learning in order also to clarify the usage and advantage for each educational technology.

A. E-learning VS. M-learning

It can be realized that learning through computer usually defined as E-learning and learning anytime and everywhere through mobile phone, smart phone normally defined as M-learning.

Table I compares between E-learning and M-learning by

focusing on what could be the advantage for E-learning and disadvantage for M-learning.

TABLE I: Comparison between M-learning and E-learning/ part 1

Criteria	E-learning	M-learning
Connectivity cost	Less expensive cable/wire line Local Area Network (LAN)	High expensive /Wi-Fi Mobile device
Input/Output capability	More user friendly due the use of keyboard and big screen	Less user friendly due the small size of screen
Processing power	High processor power and CPU speed	Less processor power and CPU speed

Table II also compares between E-learning and M-learning in positive viewpoint for M-learning rather than E-learning. Both E-learning and M-learning have advantage in some criteria but also have disadvantage in other criteria[10]. In general, both of them are very advantage in the area of education.

TABLE II: Comparison between M-learning and E-learning/ part 2

Criteria	M-learning	E-learning
Freedom of learning	Can be done anytime and anywhere	Cannot be done anytime and anywhere
Location director	Can use the GPS facilities	Cannot use GPS facilities
Portability	Portable device	Not portable device
Flexibility	Flexible	Not flexible

B. M-learning VS. Virtual Reality and Augmented Reality

Few studies were conducted to show the impacts of Augmented Reality, Virtual learning environment and mobile learning in education. In the past decade, emergent technologies such as Augmented Reality, Virtual learning environment and mobile learning has been on the rise to improve the learning experience in different learning environment [33].
Table III illustrates the difference between Augmented and Virtual Reality and M-learning.

Those new trends in education have different strength to improve the teaching and learning environment. Furthermore, they increase the motivation, collaboration, engagement and interest. However, in some cases mobile learning is slightly better than Augmented Reality and Virtual learning environment. This is because M-learning can help both students and teachers to conduct more types of teaching and learning pattern comparative to Augmented Reality and Virtual learning environment through the advancement of mobile technologies. While in integration Virtual and Augmented Reality is a good selection for some subjects that require the students to visualize and also in the fields of study such as Medicine, Chemistry, Physics, Mathematics, Astronomy, History and Biology.

TABLE III: Comparison among Augmented, Virtual Reality and Mobile Learning

Criteria	Augmented Reality	Virtual Reality	M-learning
Equipment	Mobile device and flashcard	Computer	Mobile device
Mobility	Movable	Static	Movable
Place	Anywhere and anytime	Classroom or lab	Anywhere and anytime
Interaction	Many users	Many users	Many users
Connectivity	Stand-alone	On line	On line and Standalone
Display	3D	3D	2D

IV. Conclusion and Future Work

The swift growth in mobile technology and usage in many directions has been used as an effective way of learning and also to break the restriction of time and space in the traditional classroom learning.
This paper discussed the new trends in education based on different technologies to facilitate teaching and learning which are Virtual Reality, Augmented Reality, M-learning and E-learning. Furthermore, this paper discussed some criteria, advantages and disadvantages for them in a comparative study. The presented comparison could benefit the teachers who are deciding to integrate technology in the teaching and learning process by increasing students motivation, engagement, interaction and easy communication between students and teachers. As a future work, we have noticed that Augmented Reality is still new in education and still some usage limitation. Moreover, very less researches are related to the new technology called Mobile Augmented Reality(MAR) which is integrated mobile application with Augmented Reality.

References

[1] A. T. Korucu and A. Alkan, "Differences between m-learning (mobile learning) and e-learning, basic terminology and usage of m-learning in education," *Procedia - Social and Behavioral Sciences*, vol. 15, pp. 1925–1930, 2011. [Online]. Available: http://dx.doi.org/10.1016/j.sbspro.2011.04.029

[2] T. Georgiev, E. Georgieva, and A. Smrikarov, "M-learning-a new stage of ?-learning," in *International conference on computer systems and technologies-CompSysTech*, vol. 4, no. 28, 2004, pp. 1–4.

[3] N. Selwyn, *Education and technology: Key issues and debates*. Bloomsbury Publishing, 2016.

[4] M. H. A. Rahman, N. Yahaya, and N. D. A. Halim, "Virtual world for collaborative learning: A review," in *Teaching and Learning in Computing and Engineering (LaTiCE), 2014 International Conference on*. IEEE, 2014, pp. 52–57.

[5] G. Attwell, "Personal Learning Environments - the future of eLearning ?" *eLearning Papers*, vol. 2, no. January, pp. 1–8, 2007.

[6] X. Huang, "An E-learning System Architecture based on Cloud Computing," vol. 6, no. 2, pp. 74–78, 2012.

[7] K. Nyíri, "Towards a philosophy of m-learning," in *Wireless and Mobile Technologies in Education, 2002. Proceedings. IEEE International Workshop on*. IEEE, 2002, pp. 121–124.

[8] F. Ozdamli and N. Cavus, "Basic elements and characteristics of mobile learning," *Procedia-Social and Behavioral Sciences*, vol. 28, pp. 937–942, 2011.

[9] J. E. D. Vera, *Left to my own devices: Learner autonomy and mobile-assisted language learning*. Brill, 2012, vol. 6.

[10] N. Y. Asabere, "Towards a Perspective of Information and Communication Technology (ICT) in Education : Migrating From Electronic Learning (E-Learning) to Mobile Learning (M-Learning)," *International Journal of Information and Communication Technology Research*, vol. 2, no. 8, pp. 646–649, 2012.

[11] L. Zhong, Q. Xue, H. Liu, and J. Yuan, "Research on 3g mobile learning based on cloud service," in *E-Product E-Service and E-Entertainment (ICEEE), 2010 International Conference on*. IEEE, 2010, pp. 1–4.

[12] A. Kukulska-Hulme, "Cognitive, ergonomic and affective aspects of pda use for learning," 2002.

[13] S. Bidin and A. A. Ziden, "Adoption and Application of Mobile Learning in the Education Industry," *Procedia - Social and Behavioral Sciences*, vol. 90, no. InCULT 2012, pp. 720–729, 2013. [Online]. Available: http://linkinghub.elsevier.com/retrieve/pii/S1877042813020338

[14] J. Preece, *Online communities: Designing usability and supporting socialbilty*. John Wiley & Sons, Inc., 2000.

[15] N. Kiyanovska, N. Rashevska, and V. Tkachuk, "Mobile technologies in education," *Metallurgical and mining industry*, no. 5, pp. 40–43, 2016.

[16] D. Cortiz and J. O. Silva, "Web and virtual reality as platforms to improve online education experiences," in *Human System Interactions (HSI), 2017 10th International Conference on*. IEEE, 2017, pp. 83–87.

[17] C. Jennett, A. L. Cox, P. Cairns, S. Dhoparee, A. Epps, T. Tijs, and A. Walton, "Measuring and defining the experience of immersion in games," *International journal of human-computer studies*, vol. 66, no. 9, pp. 641–661, 2008.

[18] M. Zyda, "From visual simulation to virtual reality to games," *Computer*, vol. 38, no. 9, pp. 25–32, 2005.

[19] J. Martín-Gutiérrez, C. E. Mora, B. Añorbe-Díaz, and A. González-Marrero, "Virtual technologies trends in education," *EURASIA Journal of Mathematics Science and Technology Education*, vol. 13, no. 2, pp. 469–486, 2017.

[20] X. Du and A. Arya, "Design and evaluation of a learning assistant system with optical head-mounted display (ohmd)," in *International Conference on Learning and Collaboration Technologies*. Springer, 2015, pp. 75–86.

[21] D. Griol, J. M. Molina, and Z. Callejas, "An approach to develop intelligent learning environments by means of immersive virtual worlds," *Journal of Ambient Intelligence and Smart Environments*, vol. 6, no. 2, pp. 237–255, 2014.

[22] A. S. Mathur, "Low cost virtual reality for medical training," in *Virtual Reality (VR), 2015 IEEE*. IEEE, 2015, pp. 345–346.

[23] L. Madden, *Professional augmented reality browsers for smartphones: programming for junaio, layar and wikitude*. John Wiley & Sons, 2011.

[24] P. S. Medicherla, G. Chang, and P. Morreale, "Visualization for increased understanding and learning using augmented reality," in *Proceedings of the international conference on Multimedia information retrieval*. ACM, 2010, pp. 441–444.

[25] J. Carmigniani, B. Furht, M. Anisetti, P. Ceravolo, E. Damiani, and M. Ivkovic, "Augmented reality technologies, systems and applications," *Multimedia Tools and Applications*, vol. 51, no. 1, pp. 341–377, 2011.

[26] T. Mazuryk and M. Gervautz, "Virtual reality-history, applications, technology and future," 1996.

[27] N. F. Saidin, N. D. A. Halim, and N. Yahaya, "A review of research on augmented reality in education: Advantages and applications," *International education studies*, vol. 8, no. 13, p. 1, 2015.

[28] S. C.-Y. Yuen, G. Yaoyuneyong, and E. Johnson, "Augmented reality: An overview and five directions for ar in education," *Journal of Educational Technology Development and Exchange (JETDE)*, vol. 4, no. 1, p. 11, 2011.

[29] S. Singhal, S. Bagga, P. Goyal, and V. Saxena, "Augmented chemistry: Interactive education system," *International Journal of Computer Applications*, vol. 49, no. 15, 2012.

[30] C. Coffin, S. Bostandjiev, J. Ford, and T. Hollerer, "Enhancing classroom and distance learning through augmented reality," in *EdMedia: World Conference on Educational Media and Technology*. Association for the Advancement of Computing in Education (AACE), 2010, pp. 1140–1147.

[31] E. P. Burton, W. Frazier, L. Annetta, R. Lamb, R. Cheng, and M. Chmiel, "Modeling augmented reality games with preservice elementary and secondary science teachers," *Journal of Technology and Teacher Education*, vol. 19, no. 3, pp. 303–329, 2011.

[32] E. Medina, Y.-C. Chen, and S. Weghorst, "Understanding biochemistry with augmented reality," in *EdMedia: World Conference on Educational Media and Technology*. Association for the Advancement of Computing in Education (AACE), 2007, pp. 4235–4239.

[33] J. Bacca, S. Baldiris, R. Fabregat, S. Graf *et al.*, "Augmented reality trends in education: a systematic review of research and applications," *Journal of Educational Technology & Society*, vol. 17, no. 4, p. 133, 2014.

A STUDY ON INDIVIDUAL AND ORGANISATIONAL LEARNING INFLUENCE ON SUCCESS AMONG MICRO, SMALL AND MEDIUM ENTREPRENEURS(MSMEs)

Dr.V.Ramanujam*, Dr.T.Viswanathan** and Dr.K.Lingaraja***

*Assistant Professor, Bharathiar School of Management and Entrepreneurs Development, Bharathair University, Coimbatore – 641 046 Tamil Nadu, India Email: drvramanujam07@gmail.com

**Assistant professor, Symbiosis Institute of Business Management, Bangalore – 560 100 Karnataka, India. Email: viswanathan@sibm.edu.in

***Faculty of Management, Bharathiar School of Management and Entrepreneurs Development, Bharathair University, Coimbatore – 641 046 Tamil Nadu, India
Email: klingarajaphd@gmail.com

Abstract

Many new businesses emerge every year of which few survive and many fail. What leads to the success of Entrepreneurs? Researchers, intrigued by the question have explored the factors that influences entrepreneurial success. The major factors of success identified in the past are Character ethics, Professional ethics, Individual and Organizational learning. In an established business enterprise, the decision of Intrapreneurs are largely influenced by causal reasoning. They look for effective means among the alternatives to achieve a predetermined goal. Whereas in a small business, the Entrepreneurial decisions are based on effectual reasoning. They try with various means and allow the goal to emerge contingently over time. Therefore, every means leads to a process of learning whether the outcome is a success or failure. Thus the individual learning leads to organizational learning. Organizations that are able to institutionalize the successful means through processes and addressed the failures by revalidating the assumptions have emerged as learning organizations. So among other factors, learning behavior in an entrepreneur is considered to be one of the major root cause for entrepreneurial success. Therefore, entrepreneurial learning has emerged as a crucial important concept to be studied. The main aim of this research is to examine whether individual and organizational learning influences entrepreneurial success. Data were collected from 150 Micro small medium entrepreneurs in Coimbatore district, state of Tamilnadu in India. The study reveals that there is a positive relationship between individual and entrepreneurial learning vis-à-vis the success of small and medium enterprises.

Key words: Individual learning, Organizational learning, Organizational success, Knowledge, Experience

1. INTRODUCTION

The success of a Micro, Small and Medium Enterprise (MSMEs) depends on the nature of its business and the entrepreneurial characteristics. Among various factors leading to the success of SMEs, individual and organization learning are the major constituents of success. Learning is considered as the critical competency that determine success (Dixon 1992). Learning is the process whereby knowledge is created through the transformation of experience. (Kolb 1984). Organizational learning (OL) is a crucial concept for organizational improvement in raising the performance and widely for competitive value (Friedman et al., 2005). The entrepreneurs of MSMEs enjoy freedom in making decisions. While some of the managerial decisions contribute to the success and few may lead to the causes for failure of the organization. The reflection on success and failures creates the foundation for experiential learning in an organization. The organizational learning goes through a process of Institution, Interpretation, Integration and Institutionalization. Thus individual and organization learning plays a pivot role in determining entrepreneurial success. The organization learning takes place in two ways (a) self-learning through the experience of its members and (b) dissemination of new knowledge acquired by other members of the organization. The relationship between the individual learning and organizational learning is still remained as the unexplored issue for the researchers. This gap is filled by the current study that examines the relationship between individual and organizational learning and its influence on the success of organization. In this study, we have constructed a framework based on the 4I framework of organizational learning (Marry M Crossman, et al 1999) and the 3-P model Tynjälä (2013) to determine whether individual and organizational learning influences Entrepreneurial success.

2. REVIEW OF LITERATURE

The "mental models" are accepted to be pivotal components in the delegate procedure of making an interpretation of individual learning into hierarchical learning **(Kim, 1993)** and has been recognized by various researchers in their significance field (e.g. **Hodgkinson and Sparrow,2002; Kim, 1993**;). Associations will be more successful when significant segments, (for example, structure, innovation, frameworks, individuals, and culture) are compatible with each other and when there is a solid match between the association and its condition (**Nadler, 1987**). Organisational learning is argued to be integral for achieving and maintaining this fit **(Boland et al., 1996; Dougherty, 1992)**.

Buchanan and Huczynski (2010) guarantee that individual learning hypothesis can be isolated into the behavioural and the intellectual methodologies. These two methodologies expect that experience influences conduct and that people can't learn without input. Thus 'criticism' might be an empowering agent of individual and authoritative learning.

Crossan, Lane and White (1999) distinguish and represent the connection amongst individual and hierarchical learning. The 4I structure is imperative with a specific end goal to comprehend institutional components that impact learning activity and process, and it accordingly demonstrates how individual and hierarchical learning rely upon each other.

Katharina (2014)investigate the idea of individual learning in enabling organizational change and therefore toprovide a framework of key barriers and enablers of learning. This study aims to analyze how individual learning approaches are applied in small and medium-sized enterprises (SMEs) in Germany and which barriers and enablers result from this.

Asindividual learning in organizations evolves into and contributes to group and organizational learning, this was considered while detecting factors that influence learning.

Sascha G(2015) purpose of this paper is to cumulate published empirical studies (1980-2009) onthe relationship between individual variables and entrepreneurial status. The authors categorize repeatedly analyzed determinants into six perspectives (trait, cognitive, affective, intentions, learning, and economic), review empirical findings for each determinant and each perspective, investigate trends in the field, and propose avenues for future research.

3. CONCEPTUAL FRAMEWORK OF THE STUDY

Organizations ultimately learn via their individuals, hence theories of individual learning are crucial for understanding OL (Kim, 1993). However, OL is much more complex and dynamic than being just a simply enlarged individual learning theory (ibid.). In this study, we draw inputs from the 4I framework of organizational learning and the 3-P model to examine the relationship between learning environment and organizational success.

Factors Identified for Measuring the Organizational Learning, Individual Learning and Organizational Success

The final scales for the organizational learning, individual learning and organizational success are presented with following number of items, 7 items measure the individual learning, 8 items measure the organizational learning and 7 items measure the organizational success. The statements are adapted with existing items, previously tested and validated scales of Watkins and Marsick (Yang et al., 2004) and Bontis et al. (2002). Yang et al., 2004 and Bontis et al. (2002) confirmed dimensions of learning organisation questionnaire and also his study revealed strong evidence of validity and reliability for the scale measuring dimensions of the learning organization. However, some changes were made to incorporate the scales. The items consider for the individual learning are taken from Bontis et al. (2002), Yang et al., (2004), Watkins and Marsick are "individuals are confident about accomplishing their work", "individuals are able to focus on what exactly is to be accomplished", "individuals take pride in their work", "feel a sense of accomplishment"," high levels of energy at work", "individuals act current and knowledgeable", "individuals are aware of critical issues". The items consider for Organisational learning are considered from Bontis et al. (2002), Yang et al., (2004) "clear cut vision mission and strategy", "organizational structure supports its strategic direction", "organizational culture is innovative", "organizational structure allows people to work effectively", "organization has built a culture of trust", "organization has developed operational procedures", "organization maintains an up to date database", "organization has developed systems to nurture knowledge management" are considered. The item considered for organisational success are "Employees have the right information to do the job", "Our employees will respond to change", "Decision are based on the employee's suggestions and consideration", "Employees adhere to ethic code", "Organization encourages the innovativeness", "organization a good professional relationship with employee is maintained". "We retain existing clients and manage to attract new ones" considered.

Figure 1. A Conceptual model measuring Individual learning and organizational learning influencing organizational success.

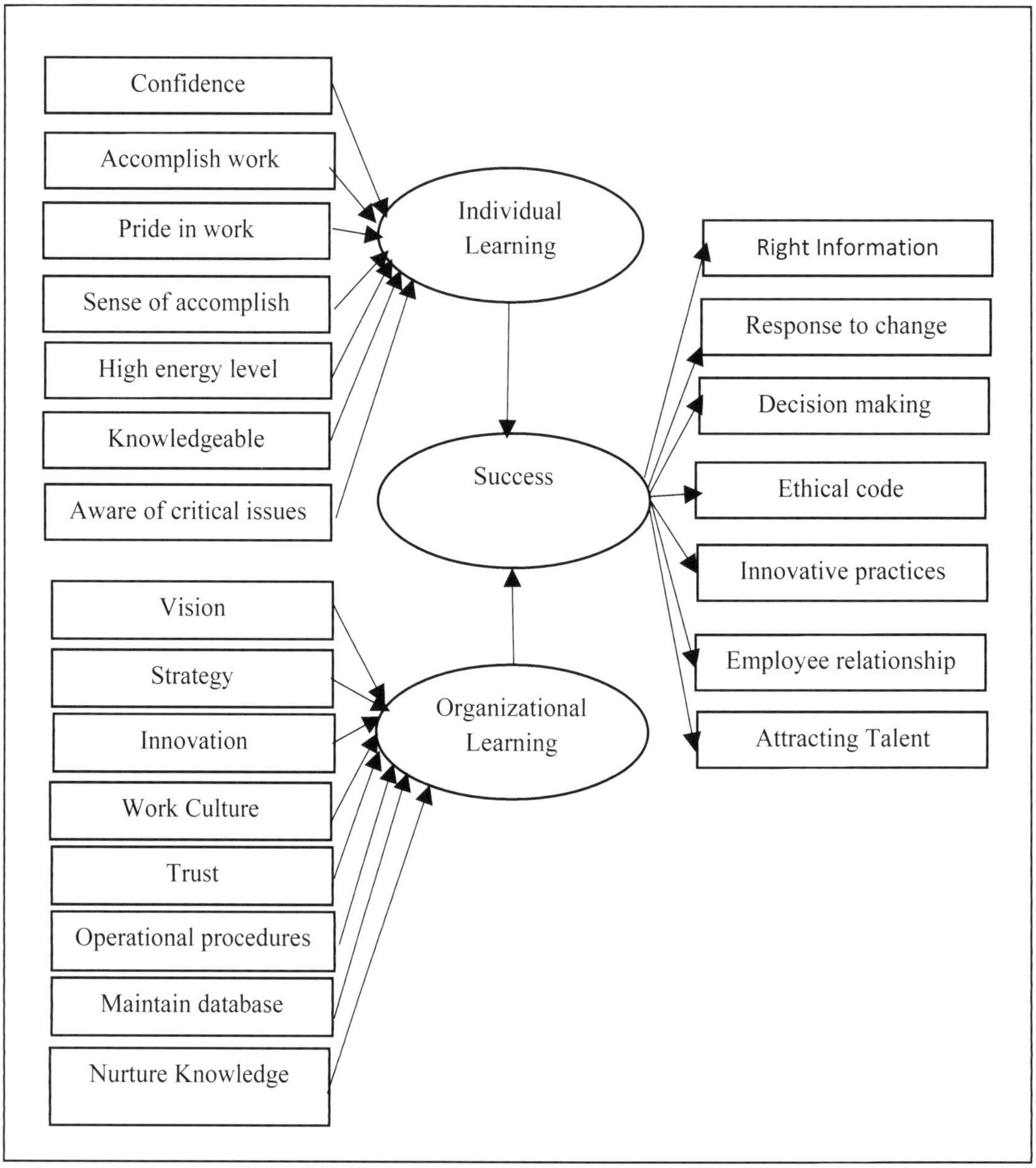

4. STATEMENT OF PROBLEM

MSMEs play a crucial role in Economic development by making significant contribution to GDP and employment generation. The sector contributes to 30 % of India's GDP and creates more than 80 million jobs. Therefore, the sustainability of these enterprises is essential for the growth of economy. Sustainability is an aspect of two factors, a) business related such as the external & internal environment and b) the characteristics, traits of Entrepreneurs and learning experience of the enterprise. In addition, the dimensions of sustainability can be looked in terms of the means and ends of the enterprise. While the business results (the ends) are visible, the cause or the means for the ends need to examined constructively. Hence this study attempts to examine whether the traits and learning experience of entrepreneurs serve as a mean to produce the end result. Does the individual and organizational learning have an impact on the success of enterprises. A well-established firm take business decisions in a conducive environment where there is reasonable access to all necessary resources. In contrast, the Entrepreneur of a MSME is bound to take decision in an environment constrained with various challenges of external and internal environment. As a result, the organization experience various cycle of learning and experience. According to Kolb learning style, it includes Concrete experience, Reflective observation, Abstract conceptualization and Active experimentation. Thus learning plays a major role as a means to success. This study aims to study the relationship between organizational learning and its influence on entrepreneurial success.

5. RESEARCH QUESTION

This study addresses the question whether individual and organizational learning have an influence on the overall success of the Micro Small and Medium Enterprises.

6. OBJECTIVE OF STUDY

- To measure the relationship between organization and individual learning.
- To study the variables effecting the organizational success.
- To examine the impact of individual and organization learning on the organizational success.

7. SCOPE OF THE STUDY

Coimbatore is a major commercial and business hub in the state of Tamil Nadu. The business tradition of Coimbatore started when the first cotton mill was set up in 1888. This has led to the growth of cotton cleaning and pressing factories and was exported to Bombay and England. The city hosts 15 % of textile mills in India and later known as "The Manchester of India". The progressive growth of cotton mills in and around the city created the need for Plant and Machinery, Foundries and spares. The foundries started to leverage their manufacturing capabilities to cater to the demands of supplying components to automobile and other industries. Over a period, the city emerged as a strong base for supply of auto components for the whole of India. Due to the conducive environment and entrepreneurial spirit, a number of ancillary units have evolved in the last four decades. Coimbatore is a home to more than 50,000 Micro, Small and Medium enterprises in the various fields of general engineering, textiles, textile spates, auto spares, machine tools, motor & pumps, foundry, home appliances, software, farm equipment etc., in manufacturing. A century old entrepreneurial orientation, availability of skilled labor, work culture and favorable business environment attributes for sustainable development of MSMEs in the region. It is the highest revenue yielding district in the state even ahead of Chennai, thus making it one of the fastest-growing second-tier metro cities in India. Coimbatore is ranked as one of most competitive (by business environment) Indian cities. There the district provides enormous research opportunities with respect to the growth, success and challenges of Entrepreneurs. Thus the scope of the study is to focus on the MSMES in Coimbatore district to examine the reasons for entrepreneurial success. The study concentrates on entrepreneurial characteristics and how individual and organization learning experience leads to entrepreneurial success.

8. HYPOTHESES OF STUDY

The hypotheses are formed to test whether individual and organizational learning has an effect on organizational success. The following are the hypotheses
Ho Individual learning has no impact on organizational success.
Ho Organizational learning has no impact on organizational success.

9. LIMITATION OF THE STUDY

The study examines whether the organizational and individual influences the success of enterprises. However other exogenous variables may have influence on the success. Such variables are the outcome of political, economic, societal and technology factors. Various other macro and micro economic factors may also influence the success of organization.

10. RESEARCH METHODOLOGY

The research design adopted for the study is descriptive in nature. The questionnaire consists of 22 statements that measure the influence of individual and organizational learning on organizational success. After finalizing the number of items in the research instrument, a pilot study was undertaken. The population of the study is MSME entrepreneurs of Coimbatore district. The survey was conducted by means of Stratified Random Sampling; the MSME entrepreneurs are divided into different industries (strata) so that stratified sampling is used under this study. The target population for this study consists of 980 MSME entrepreneurs who are registered with District Industrial Centre (DIC), in Coimbatore district. The sample selection was based on the criterion that the firm had to be established for at least three years because it is not possible to gather enough data to consider the performance of firms operating in a shorter period of time. Both primary and secondary sources are used for the data collection. The primary data was obtained using questionnaire methods. Finally 513 respondents questionnaires are considered to be appropriate to conduct the study. Secondary data information was obtained from the published documents of SIDCO, Tamil Nadu small and Tiny Industries Association (TANSTIA).

ANALYSIS OF INDIVIDUAL LEARNING AND ORGANISATIONAL LEARNING ON SUCCESS OF THE MSME ENTREPRENEURS

i) RESPONDENTS OPINION ON INDIVIDUAL LEARNING, ORGANISATIONAL LEARNING AND SUCCESS OF THE MSME ENTREPRENEURS

The mean score is calculated in order to obtain the respondents opinion on individual learning, organisational learning and success of the MSME entrepreneurs.

Table1. Respondents Opinion on Individual Learning, Organisational Learning and Organisational Success of the MSME Entrepreneurs

Sl.No	Statements	Mean score
1.	Individuals are confident about accomplishing their work effectively	2.6692
2.	Individuals are able to focus on what exactly is to be done to accomplish the work effectively	2.5851
3.	Individuals take pride in their work	2.7973
4.	Individuals feel a sense of accomplishment in what they do	2.8935
5.	Individuals have high levels of energy at work	3.0839
6.	Individuals act current and knowledgeable about their field of work	3.1718
7.	Individuals are aware of critical issues that affect their work	3.3886
8.	Organization has a clear cut vision mission and strategy for the future	3.5235
9.	Organizational structure supports its strategic direction.	3.6894
10.	Organizational culture is innovative.	3.4314
11.	Organizational structure allows people to work effectively	3.6453
12.	Organization has built a culture of trust among people	3.5729
13.	Organization has developed operational procedures to guide its activities and help employees and groups to work efficiently	3.5235
14.	Organization maintains an up to date database of its knowledge inventory	2.7973
15.	Organization has developed systems to nurture knowledge management	2.4935
16.	Employees have the right information to do the job	3.1149
17.	Our employees will respond to change	2.4691
18.	Decisions are based on the employee suggestions and consideration.	2.8864
19.	Employees adhere to ethic code	2.5041
20.	Organization encourages the innovativeness	3.6682
21.	A good professional relationship with employee is maintained	2.7674

From the above mean table, the respondents opinion on individual learning organizational learning and organizational success are analyzed. The highest mean score value with 3.689 score on "Organizational structure supports its strategic direction". "Organizational structure allows people to work effectively" scores 3.6453. The least score mean value is 2.4691 for the statement "Our employees will respond to change".

ii) Stage: 1 Reliability And Validity Measurement Between Individual Learning And Organisational Learning On Success:

A Partial Least Square (PLS) model is applied in two stages sequentially. First is the assessment and refinement of adequacy of the measurement model, followed by the assessment and evaluation of the structural model. This is to ensure the reliability and validity of the measures prior to the attempt in making and drawing conclusions on the structural model.

Measurement model evaluation is aimed to evaluate the consistency and validity of the dimensions. Consistency evaluations are done through individual indicators and construct reliability tests. While validity of the variables is tested based on convergent and discriminant validity, individual statements reliability explains the variance of individual statements relative to latent variable by calculating standardised outer loadings of the latent variables. Latent variables with outer loading 0.7 or higher are considered highly satisfactory. While loading value of 0.5 is regarded as acceptable, the latent variables with loading value of less than 0.5 should be dropped. Cronbach's alpha and Composite Reliability (CR) indicate how well a set of latent variables appraises a single latent construct. However, compared to Cronbach alpha, composite reliability is considered a better measure of internal consistency because it employs the standardized loadings of the latent variables. Litwin suggested that value of Cronbach alpha should be higher than 0.7 and for composite reliability, the value of 0.7 is suggested as "acceptable".

For the validity of the variable, the variables are tested on convergent and discriminant validities. Convergent validity is carried out by Average Variance Extracted (AVE) test on variables. It determines the amount of variance captured by latent variable from its relative latent variables due to measurement errors. Barclay et al. and Hair et al. argued that a minimum 50% of the variance from manifest variable should be captured by latent variables. This implies that AVE value of the construct should be greater than 0.5. Discriminant validity is carried out to confirm that the latent variable in any construct is relevant to the designated latent variable where its cross-loading value in LV is higher than that in any other constructs.

Table 2: Results Summary of the Measurement Model of success

Construct	Indicators	Loadings	Indicator Reliability	Cronbach's Alpha	Composite Reliability	AVE
Individual Learning	IL1	0.678	0.549	0.880	0.902	0.578
	IL 2	0.752	0.565			
	IL 3	0.867	0.751			
	IL 4	0.756	0.571			
	IL 5	0.712	0.506			
	IL 6	0.751	0.564			
	IL 7	0.867	0.751			
Organizational Learning	OL1	0.771	0.594	0.972	0.976	0.819
	OL2	0.876	0.767			
	OL3	0.781	0.609			
	OL4	0.879	0.772			
	OL5	0.783	0.613			
	OL6	0.891	0.793			
	OL7	0.967	0.935			
	OL8	0.933	0.870			
	SUC1	0.734	0.538			

	SUC1	0.734	0.538			
	SUC3	0.754	0.568			
	SUC4	0.901	0.811			
	SUC5	0.912	0.831			
	SUC6	0.715	0.511			
	SUC7	0.835	0.697			

The table 2 reveals that the composite reliability, cronbach alpha, AVE, loadings, and indicator loading of the individual Learning, Organizational learning and success, all the obtained values were greater than the threshold value 0.4 (Wold 1998). Hence **they are acceptable.** Squaring of the indicator loadings provide the indicator reliability values which are shown in the fourth column of Table 2. Although indicator reliability value of 0.7 or higher is preferred, values higher than 0.4 are acceptable (Clarck and Watson 1995).As all the indicator loadings in the table are greater than the threshold value of 0.4 the indicators of the constructs are accepted. The Cronbach's alpha scores for all the constructs (for individual learning 0.880, for organizational learning 0.972, for success 0.954) which areshown in column 5 of the Table 2 are above 0.7 which indicate that internal consistencies of the constructs is reliable. The composite reliability values for individual learning 0.902, for organizational learning 0.976, for success 0.960 shown in column 6 of the table are greater than the threshold value of 0.7. Therefore the set of indicators measuring the latent constructs is ensured.The above three estimations are very well greater than the threshold values. Hence it is confirmed that the constructs of the measurement model are reliable.Validity of constructs of Measurement Model 1- For validity assessment convergent validity and discriminant validity are checked. The Convergent Validity is calculated for each construct are shown in the seventh column of Table 2 areIndividual Learning (AVE = 0.578), Organizational Learning (AVE = 0.819), Success of Organization (AVE = 0.707. As the computed AVE values are greater than 0.5, the convergent validity of the constructs in the measurement model is confirmed.

Table 3: Correlation and measures of validity among the variables

	AVE	Individual Learning	Organizational Learning	Success
Individual Learning	0.578	**0.742**		
Organizational Learning	0.819	0.512	**0.908**	
Success	0.707	0.345	0.611	**0.811**

The Table no.3 shows that based on Fornell and Larcker method, the discriminant validity of Individual Learning,Organizational Learning, and Success of Organizationis proved since the square root of AVE of a construct is greater than the correlation values of that construct with other constructs.The correlation value of individual vakue(.512) Organisational learning (.611) and success (.345) when compared with square root of AVE (0.742, 0.908, and 0.811) the square root of AVE values are greater. Thereby discriminant validity is attained between Individual Learning, Organizational Learning and success of MSME entrepreneur.

Figure1: Measuring the measurement model

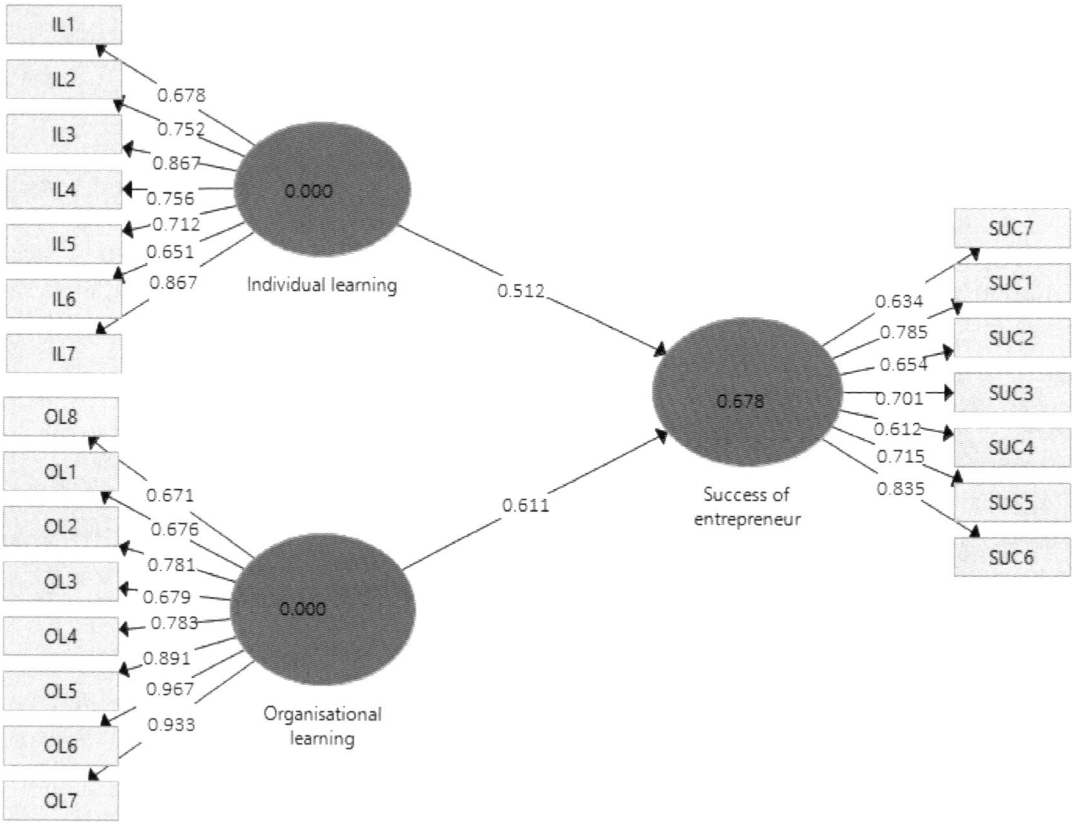

iii) Stage: 2 Model Fit For Relationship Between Individual Learning, Organizational Learning And Success Of Entrepreneur:

Structural model measures relationship between individual learning, organizational learning and successof entrepreneur latent variables through evaluating R^2 value, that is, coefficient of determination. R^2 corresponds to the degree of explained variance of endogenous latent variables. cohen et al. for a good model, the value of R^2 of endogenous latent variable should be more than 0.26.

Table 4 Obtained Coefficients of Determination (R^2)

	R Square	R value
Success	0.678	-
Individual Learning	-	.512
Organisational Learning	-	.611

From the figure, Since R^2 value for the conceptual model is higher than the suggested value, the model is considered to have substantial degree of explained variance of success by entrepreneur's individuals learning factors.The R2 in the Table 4 and figure 1 explains that 67.8% of impact in success is explained byIndividual Learning, and Organizational Learning.

Table 5Measuring the Variance Inflation Factor (VIF) of each construct

Factors	VIF
Individual Learning	2.321
Organizational Learning	2.617
Success	3.412

Note: Success is taken as dependent variable

On checking Table 5 the collinearity for measurement model that is between variables of success, where all the obtained VIF value is < 5 there is no collinearity issue between the latent variables. The VIF value for Individual Learning is 2.321, for Organizational Learning is 2.617, and Success is 3.412. Therefore all the constructs value is less than 5 and hence there is no collinearity among the constructs.

Figure 2: Model Fit for Relationship between Individual Learning, Organizational Learning and Success of Entrepreneur

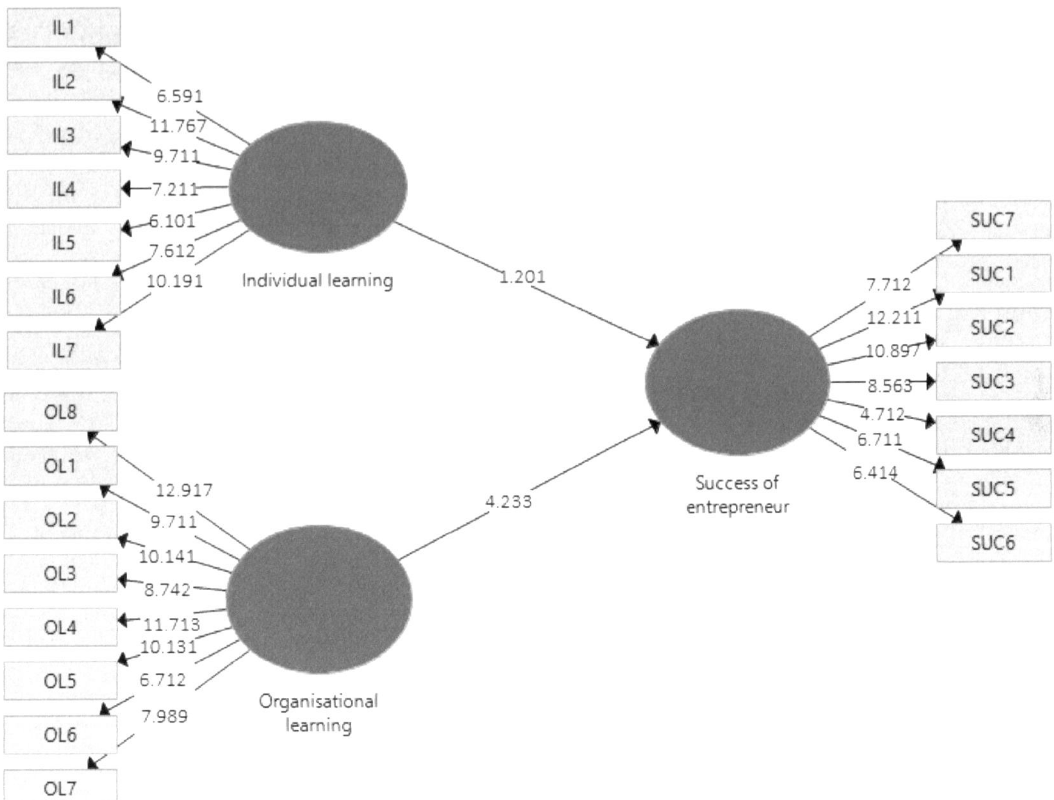

Table 6 Results of structural model

	Null Hypotheses	Paths	Original Sample	T Statistics	Results of Hypotheses
$H_0$1	Individual learning does not have impact on success.	**Individual learning-> success**	-0.622	1.201	Not Significant
$H_0$2	Organizational learning does not have impact on success	**Organization al learning-> Success**	0.012	4.233	Significant

The table 6 shows the result of the specific hypotheses predicted in this study. By performing the bootstrapping procedure in Smart PLS, *t*-values are generated for the model. The setting for the complete bootstrapping procedure included 500 subsamples, and a two-tailed t test is conducted at $\alpha = 0.05$ significant level. The evaluation criteria for confirming each hypothesis was the use of t values for each path loading. Significant t-values for path loadings signify support for the proposed Path Mean Standard Deviation T-Statistic hypothesis. The cut-off criteria according to hair et al. were a t-value greater or equal to 1.96 for an alpha level of 0.05 (Hair et al. 2006). Finally the individual learning has a positive impact on the organizational success. The Structural path coefficient value between the individual learning and success is 1.201 statistically not significant; therefore individual learning does not impact MSME entrepreneur's success. The Organisational learning has positive impact on success. The Structural path coefficient value between the Organisational learning and success is 4.233 statistically significant, Organisational learning positively impact MSME entrepreneur's success.

11. SUMMARY OF FINDINGS, SUGGESSTIONS AND CONCLUSION

Summary of findings

From the mean score table the respondents opinion on individual learning organizational learning and organizational success are analyzed. The highest mean score value with3.689scoreson "Organizational structure supports its strategic direction". "Organizational structure allows people to work effectively" scores 3.6453. The least score mean value is 2.4691 for the statement "Our employees will respond to change". From the Structural equation modeling using smartpls, in first stage measurement model the reliability and the validity is achieved, the R value for both the construct are positive and they are 0.512 and 0.612 therefore individuals learning and organisational learning have a relationship with organisational success. The R square value is 67.8% therefore both individuals learning and organisational learninginfluencing the organisational success with more than better value. In second stage the structural model the t value is greater than 1.96 which satisfies and signifies each hypothesis for each path loading. Finally the result reveals that Organizational success is greatly achieved by the Organisational learning than individual learning.From the model it is analyzed that organizational learning contributes more on the success of organization.

Suggesstions

This paper reveals the importance of individuals learning and organisational learning towards the organisational success. Most of the small and medium enterprises are run by family members. Most of the learning that place in the organisation are anectodal and there is very limited scope for instutituionalisation of individual and organisational learning. It is also possible that the an individual's learning experience is confined only to him and goes along with him when he leaves the organisation. Thus the organisation suffers from knowledge deficieny. This paper provides a conceptual framwork that shows various enablers of learning and its relevance to success. Every MSME should identify the enablers of learning and the sources from which the organsiation can capture the individual and organisation learning experience. Such experiences should be institunalised through robust process. It is also pertinent for the organisation to scan the external environment and learn from the changes. An approprite leaning management system may be estalished to leverage the individual and organisational learning in to the entrepreneurial success.

Conclusion

Finally this paper concludes that there is a positive relationship between all three latent variables individual learning,organisational learning and success. The bootstrap result reveals that t value for individual learning and success is not significant when compared to organisational learning and success. Organisational learning acts as important variable/ factor in influencing the organisational success in MSME entrepreneurs. Therefore influence of Organisational learning on success is important in enterpreneurship.

REFERENCES

Argyris C. & Schön D. A., (1996), Organizational Learning II: Theory, Method, and Practice, Boston, Reading, Mass: Addison-Wesley.

Baron shane (2008) Entrepreneurship a process perspective 2nd edition, Boston, Cengage Learning.

Boland, R., Ramkrishnan, V. and Teeni, D. (1996), "Designing information technology to support distributed cognition", in Meindl, J., Stubbart, C. and Porac, J. (Eds), Cognition within and between Organizations, Sage, London

Buchanan, D.A. and Huczynski, A. (2010) Organizational behaviour, 7th edn, Harlow, Pearson Education.

Crossan, M.M., Lane, H.W. and White, R.E. (1999) An Organizational Learning Framework: From Intuition to Institution, The Academy of Management Review, vol. 24, no. 3, pp. 522–537

Easterby-Smith, M. and Araujo, L. (1999), "Organisational learning: current debates and opportunities", Organisational Learning and the Learning Organisation: Developments in Theory and Practice, Sage, London.

Easterby-Smith, M. and Lyles, M. (Eds) (2003), Handbook of Organisational Learning and Knowledge Management, Blackwell, Oxford.

Fenwick, T. (2008). Workplace learning: emerging trends and new perspectives. New Directions for Adult and Continuing Education, 119, 17–26.

Harteis, C., Bauer, J., & Gruber, H. (2008). The culture of learning from mistakes: how employees handle mistakes in everyday work. International Journal of Educational Research, 47, 223–231.

Duncan, R. and Weiss, A. (1979), "Organisational learning: implications for organisational design", in Staw, B.M. (Ed.), Research in Organisational Behaviour, Vol. 1, JAI Press,Greenwich, CT.

Dixon, N. (1994), The Organisational Learning Cycle: How We Can Learn Collectively, McGraw-Hill, London.

Dougherty, D. (1992), "Interpretive barriers to successful product innovation in large firms", Organisation Science, Vol. 3 No. 2, pp. 179-202.

Friedman, V., Lipshitz, R. and Popper, M. (2005), "The mystification of organisational learning", Journal of Management Inquiry, Vol. 14 No. 1, pp. 19-30.

Sascha G. Walter Simon Heinrichs , (2015),"Who becomes an entrepreneur? A 30-years-review ofindividual-level research", Journal of Small Business and Enterprise Development, Vol. 22 Iss 2 pp, 225 – 248.

Salojärvi, S., Tynjälä, P., Myyry, L., Ikonen-Varila, M., & Nikkanen, P. (2010). How can a learning networksupport organisational development? International Journal of Strategic Change Management, 2(4),375–395.

Slotte, V., Tynjälä, P., & Hytönen, T. (2004). How do HRD practitioners describe learning at work? HumanResource Development International, 7(4), 481–499.

Schwartz, S. H. (2011). Studying values: personal adventure, future directions. Journal of Cross-Cultural

Simons, R.-J., Germans, J., & Ruijters, M. (2003). Forum for organisational learning: combining learning atwork, organisational learning and training in new ways. Journal of European Industrial Training, 27(1), 41–48.

Spector, J. and Davidsen, P. (2006), "How can organisational learning be modelled andmeasured?", Evaluation and Program Planning, Vol. 29 No. 1, pp. 63-69.

Hendry, C., Jones, A. M., & Cooper, N. Creating a learning organization: Strategies for change. Sutton Coldfield: Man-Made Fibers Industry Training Organization, 1994.

Hodgkinson, G. P., & Sparrow, P. R. (2002). The Competent Organization: A Psychological Analysis of the Strategic Management Process. Maidenhead: Open University Press.

Hair, J., Black, W., Babin, B., Anderson, R., & Tatham, R. (2006). Multivariate data analysis (6th ed.). Uppersaddle River, N.J.: Pearson Prentice Hall.

Illeris, K. (2011). Workplaces and learning. In M. Malloch, L. Cairns, K. Evans, & B. O'Connor (Eds.), The
SAGE handbook of workplace learning (pp. 32–45).

Jones,O.& Macpherson, A. (2006) "Inter- Organizational Learning and Strategic Renewal in SMEs: Extending the 4I Framework", Long Range Planning, Volume 39, Issue 2, April 2006, Pages 155-175

Kayes, D.C. (2002), "Experiential learning and its critics: preserving the role of experience in management education", Academy of Management Learning and Education, Vol. 1 No. 2, pp. 137-149.

Katharina(2013), "An Investigation of the Idea of Individual Learning in EnablingOrganizational Change"Social and Behavioral Sciences 130 (2014) 247 – 256.

Kim, D.H. (1993) The Link between Individual and Organizational Learning, Sloan Management Review, vol. 35, no. 1, pp. 37-50

Kolb, D.A. and Kolb, A.Y. (2005), "Learning styles and learning spaces: enhancing experiential learning in higher education", Academy of Management Learning and Education, Vol. 4No. 2, pp. 193-212.

Lehesvirta, T. (2004). Learning processes in a work organization. Journal of Workplace Learning, 16(1/2),92–100.

Nadler, D. (1987), "The effective management of organisational change", in Lorsch, J. (Ed.),Handbook of Organisational Behaviour, Prentice Hall, Englewood Cliffs, NJ.

Probst, G.J. and Buchel, B.S. (1997), Organisational Learning: The Competitive Advantage of the Future, Prentice Hall Europe, London.

Skerlavaj, M., Stemberger, M., Skrinjar, R. and Dimovski, V. (2007), "Organisational learning culture – the missing link between business process change and organisational performance", Journal of Production Economics, Vol. 106 No. 2, pp. 346-367.

Tynjälä, P. (2010). Workplace learning in transformation. Keynote presentation at the EARLI Learning and Professional Development SIG Conference (pp. 25–27). Germany: Munich. August, 2010.

Van Gelderen, M., Brand, M., van Praag, M., Bodewes, W., Poutsma, E., & van Gils, A. (2008). Explaining entrepreneurial intentions by means of the theory of planned behaviour. Career Development International, 13(6), 538–559

Yang, B., Watkins, K.E. and Marsick, V.J. (2004), "The construct of the learning organization:dimensions, measurement, and validation", Human Resources Development Quarterly,Vol. 15 No. 1, pp. 31-55.

Yeo, R. (2002), "Learning within organisations: linking the theoretical and empirical perspectives", Journal of Workplace Learning, Vol. 14 No. 3, pp. 109-122.

****

BUSINESS ENVIRONMENT FACTORS ON BUSINESS PERFORMACE OF SMES IN TAMILNADU

K.AMMANI AMMAL
Lecturer, Department of Business studies
Ibri college of Technology
Ibri, Sultanate of Oman

DR.K.MALARMATHI
Professor, BSMED
Bharathiar University
Coimbatore, Tamil Nadu, India

Abstract

The small and medium enterprises (SMEs) play a critical role in the development of the country. The business performance of SMEs influenced by various factors. This study examines the business environment factors which influence the business performance of SMEs in Coimbatore district. This study carried out among 100 entrepreneurs in manufacturing sectors. Responses of the subjects were collected through a structured questionnaire. The hypothesis of the study tested with SPSS software. Multiple Regression analysis was used to test the relationship between environment factors and the business performance of SMEs. This study concludes that there is a strong relationship between environmental factors and the business performance of SMEs in the study area. These factors include both macro and the micro environmental factors found to be the most significant factors contributing to the business performance of SMEs.

Keywords: Business Performance, Business environment, Factors, SMEs Performance.

I. INTRODUCTION

In Indian economy, Micro, Small and Medium Enterprises sector has emerged as a highly vibrant and dynamic sector. MSMEs not only play an important role in providing huge employment oppourtunities at relatively lower capital investment than large scale industries but also support in industrialisation of rural and backward areas, thereby promising more equitable distribution of national income and wealth by reducing regional imbalances in the country. MSMEs are complementary to large scale industries as ancillary units and this sector extremely contribute to the socio-economic development of the country. Presently, Coimbatore district is having more than 50,000 MSMEs in the various fields of manufacturing sectors such as general engineering, textiles, spares, auto spares etc., Coimbatore is ranked as one of the most competitive Indian cities, make in India also focuses on electrical machinery and electronic system as one of the promising sectors for exports from Coimbatore district. Survival and success of SMEs is not a simple phenomenon. There are several factors increase their influence and causing a huge effect on the success of SMEs (Cacciotti & Hayton, 2015). It is evident that obstacles in the business success are far more than it was previously. The environmental factors as well as some other factors are very complex and dynamic. Personal decision to become an entrepreneur is not only personal factor issues, but also environment factor issues as well (Bird, 1988). It is acknowledged that there are social, economic and social imperatives for supporting SMEs to survive and develop, particularly to support job growth. However, it is also a complex and multidimensional phenomenon. Business performance can be characterised as the firm's ability to create acceptable outcomes and actions. In order to strengthen the economy many governments in the world have been paying a more attention to development of SMEs (eg. chasten, 1992; mulhern, 1996; patrianila,2003). It is imperative to find out the influence of environmental factors in the business performance of SMEs.

II. LITERATURE REVIEW

The focus of recent studies about entrepreneurship has gradually shifted from sole personality characteristics of entrepreneurs. Research into the relationship between the environmental factors and business start-up is a more recent event. While not denying the role played by the founder's characteristics, the demand perspective suggests that the environment is more significant in understanding the business start-up. The business environment plays a vital role in the growth of SMEs (Beck and Demirguc-Kurt, 2006; Delmar & Wiklund, 2008; Stevenson and Jarill,1990). Growth of SMEs positively and negatively influence by the changes in the business environment (World Bank, 2006; Zhang, Van Doom &Leeflang, 2014). Lescevica (2002) identified that both internal and external factors are significant determinants of success among entrepreneurs. The business environment is demarcated as factors both inside and outside the organization, influencing the continued and successful survival of the organization. The internal environment includes factors in the environment that are largely controllable by the business (Fatoki & Garwe,2010; Kolstad & Wiig,2015). External environmental conditions would be associated to satisfactory government support (Yousuf, 1995), inflation, interest rates and unemployment (Van Eeden, Viviers & Venter, 2003), support of the family and friends, access to capital influence the success of SMEs (Benzing, Chu and Kara 2009). Many entrepreneurial researches agreed that one of the major problem faced by the SMEs in the developing countries is lack of financial resources (Dalberg, 2011; Cook, 2001; Grey et al 1997). Political and economic environment instability, complex taxation, poor law and order situation, and corruption are the most common problems faced by the developing countries. Consequently, in majority of the developing countries the small and medium businesses have to face a number of problems due to similar reasons (Benzing et al 2009). Specht (1993) differentiates five main environmental factors affecting organisation formation, viz., economic, political, infrastructure development, socio and market emergence factors. Fatoki & Garwe (2010) found the environment factors as economic variables and markets, corruption, labour, infrastructure and regulations. Businesses have to make decisions which deal not only with business survival opportunities, but also with business development in a changing environment under dynamic competitive conditions where each competitor tries to do impossible things to survive (Scarborough et al., 2009). Over the years, competition among SMEs has increased drastically. Competition and sustainability for SMEs involve factors such as changing market trends, changing technologies and emerging new management and organizational techniques. Stefanovic, Damnjanovic and Jasko (2010) stated that there are eight factors have impact an organisational operations likely state regulation, resource availability, supplier relationships, socio cultural structure of population demand for products/ services, sale oppourtunities and demand satisfaction capacities.

Several of the existing literature, including Salesh & Ndubisi (2006); Samad (2007); Abu Bakar et al. (2006); Aris (2006); Harvie (2004); Wang (2003); Stuti (2005); Wafa et al. (2005); Ritchie & Brindley (2000); Decker et al. (2006); Foon (2006); and SMIDEC (2007) highlight in a globalised environment SMEs are facing various challenges, for example from difficulty in facing recession, barrier from global sourcing, low productivity, lack of managerial capabilities, lack of financing, difficulty in accessing management and technology, heavy regulatory burden and others. Given the global market, SMEs are facing the lack of knowledge on legal matters. Based on the previous studies for the purpose of the study the researcher classified the environmental factors as micro and macro environmental factors that influence the business performance of SMEs.

OBJECTIVES OF THE STUDY

1. To analyse the socio-economic and demographic factors of SME entrepreneurs.
2. To identify the factors that influences the SMEs business performance in Coimbatore District.

III. RESEARCH METHODOLOGY

The study conducted among the small and medium manufacturing enterprises, which were established during the past 10 years in the Coimbatore district. Simple random sampling technique adopted to collect the data from 100 sample entrepreneurs through a structured questionnaire for the purpose of this research. The structured questionnaire of this study consists of three parts, viz., Sample entrepreneur's demographic information and the details about the firm were sought in the first part of the questionnaire which intended to measure the socio economic background of entrepreneurs and firms characteristics. The second part includes both micro environmental factors and macro environmental factors like consumers, suppliers, intermediaries, competitors, public, employees, economic factors, political and legal factors, socio-cultural factors, technological factors influence the business performance of SMEs, and the third part consists of questions to measure SMEs business performance. Likert's Five point scale anchored by strongly agree-5 and strongly disagree-1 were applied to measure the business performance of SMEs. The study used multiple regression analysis, before applying the regression analysis; the reliability test was used to test the research instrument using the values of Cronbach's alpha. To identify the relationship between environmental factors and business performance of SMEs and also used to find out the significant factors that influence the business performance of SMEs in the study area multiple regression analysis was applied.

IV. DATA ANALYSIS AND FINDINGS

The Reliability analysis was conducted to measure the consistency and the internal stability of the data used for the study. The value of Cronbach's Alpha determining the inter-item consistency and reliability measure the coefficient that reflects how well items in a set are positively correlated to one another. If the Cronbach's alpha value less than 0.5 normally considered to be low reliability, value between 0.5 and 0.7 were considered to be quite high reliability, 0.7 to 0.9 have high reliability whereas more than 0.9 considered to be very high reliability. The Table.1 shows the reliability analysis results. The Cronbach's Alpha of the items was more than 0.7 for each parts and the overall reliability was 0.850; therefore, we can infer that the structured questionnaire used in this study has both validity and reliability.

Table.1. Reliability Analysis

Variables	*Number of Items*	*Cronbach's Alpha*
Overall Reliability	75	.850
Macro Environment Factors	32	.833
Micro Environment Factors	29	.654
Business Performance	14	.706

Table 2 Demographic analysis illustrates that out of 100 sample respondents, majority of the respondents are male than the female respondents. 94% were male and the remaining 6% were female. Majority of the respondents were between the age group of 36 and 45 years, i.e., 48 %, 32% were aged above 45 years old and 50% of the respondents were graduated, 26% were from Diploma, 74% of the respondents were with previous experience related to their present business, 86% and 82% of the respondents were not acquired any training neither before start up nor after start up. On the other hand about firm analysis includes majority 58% were sole proprietorship, 84% were small enterprises, most of the enterprises were operating above 6 years less than 10 years and 61% of the enterprises having only less than 10 workers in their enterprises.

Table 2. Demographic analysis of respondents

Variable		*Frequency*	*Percent*
Age	26 - 35 years	20	20.0
	36 - 45 years	48	48.0
	Above 45 years	32	32.0
	Total	100	100.0
Gender	Male	94	94.0
	Female	6	6.0
	Total	100	100.0
Educational qualification	Post Graduate	13	13.0
	Graduate	50	50.0
	Diploma	26	26.0
	Higher secondary	8	8.0
	Primary school	3	3.0
	Total	100	100.0
Number of Years' Experience	2 years or less	10	10.0
	3 to5 years	19	19.0
	6 to10 years	36	36.0
	10 years or more	35	35.0
	Total	100	100.0
Experience related to the present business	Yes	74	74.0
	No	26	26.0
	Total	100	100.0
Training attended before starting this business	Management	10	10.0
	Technical	4	4.0
	None of them	86	86.0
	Total	100	100.0
Training attended After Starting this business	Management	2	2.0
	Technical	16	16.0
	None of them	82	82.0
	Total	100	100.0
Type of Ownership	Sole-proprietorship	58	58.0
	Partnership	26	26.0
	Private Limited	16	16.0
	Total	100	100.0

Type of Enterprise	Small scale	84	84.0
	Medium scale	16	16.0
	Total	100	100.0
Present age of the concern	Less than 2 years	10	10.0
	2 - 4 years	8	8.0
	4 - 6 years	6	6.0
	Above 6 years	76	76.0
	Total	100	100.0

Multiple Regression Analysis used to determine whether the independent variables such as macro and micro environmental factors have any significant influence on business performance of SMEs in Coimbatore District. The overall results of the regression analysis displays that this model is well constructed and it is well represented as reflected in the variables selected. Table 3, shows the multiple linear regression model summary and overall fit statistics. It was found that the adjusted R^2 of our model is .527 with the R^2 is 69 percent. This means the variables that are included in the study, can explain 69 percent variations in the business performance of SMEs in Coimbatore district. The Durbin-Watson statistic displays that the serial correlation of residuals is 2.073, the value falls within the acceptance range of between 1.5 and 2.5. Therefore, we can infer that there is no first order linear auto correlation problem in the data. Table 4 Anova explains that there is a significant influence of environment factors on business performance of SMEs. The Table 5 coefficient explains that there is no multi collinearity problem in the regression model used for this study, because the Variance Inflation Factors (VIF) and tolerance all fall within the acceptable range (VIF = 1 - 10, tolerance = 0.1 – 1.0). Lastly, to check for normality of residuals with a normal P-P plot, the plot displays that the points usually follow the normal (diagonal) line with no strong deviations. This shows that the residuals are normally distributed.

Table 3. Model Summary[b]

Model	R	R Square	Adjusted R Square	Standard Error of the Estimate	Durbin-Watson
1	.831[a]	.690	.527	1.98856	2.073

a. Predictors; (Constant), EMPLOYEES, ECONOMIC, INTERMEDIARIES, PUBLIC, CUTOMERS, SOCIOCULTURAL, TECHNICAL, COMPETITORS, SUPPLIERS, LEGAL

b. Dependent Variable: SMEBUSINESSPERFORMANCE

Table 4. ANOVA[b]

Model		Sum of Squares	df	Mean Square	F	Sig.
1	Regression	167.468	10	16.747	4.235	.003[a]
	Residual	75.133	19	3.954		
	Total	242.601	29			

a. Predictors: (Constant), EMPLOYEES, ECONOMIC, INTERMEDIARIES, PUBLIC, CUTOMERS, SOCIOCULTURAL, TECHNICAL, COMPETITORS, SUPPLIERS, LEGAL

b. Dependent Variable: SMEBUSINESSPERFORMANCE

Table 5. Coefficients[a]

Model		Unstandardized Coefficients		Standardized Coefficients			Collinearity Statistics	
		B	Std. Error	Beta	t	Sig.	Tolerance	VIF
1	(Constant)	9.069	12.419		.730	.474		
	ECONOMIC	.381	.166	.494	2.304	.033	.354	2.826
	LEGAL	-.270	.150	-.683	-1.802	.087	.113	8.823
	TECHNICAL	-.270	.310	-.243	-.872	.394	.210	4.771
	SOCIOCULTURAL	.262	.284	.260	.922	.368	.204	4.891
	CUTOMERS	-.178	.264	-.154	-.674	.509	.311	3.214
	SUPPLIERS	-.491	.335	-.362	-1.464	.160	.266	3.758
	COMPETITORS	-.469	.217	-.442	-2.164	.043	.391	2.560
	PUBLIC	.779	.246	.859	3.162	.005	.221	4.531
	INTERMEDIARIES	.449	.179	.579	2.505	.022	.305	3.282
	EMPLOYEES	.852	.238	.671	3.586	.002	.465	2.149

a. Dependent Variable: SMEBUSINESSPERFORMANCE

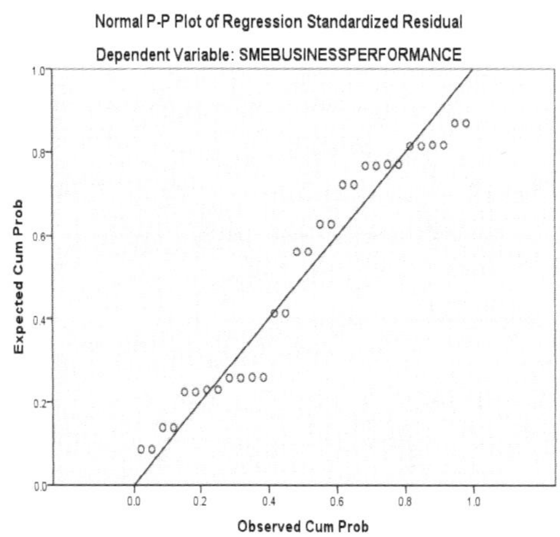

V. CONCLUSION

This research conducted to identify the factors that influence the business performance of SMEs in Coimbatore District. Analyses on Reliability and validity have demonstrated that the structured questionnaire used in this study is reliable and a valid measure. The findings of this study revealed that economic factors in the macro environment and competitors, intermediaries, employees and public of micro environment factors have significant influence on the business performance of SMEs. Political and legal factors, socio-cultural factors, technological factors, customers and suppliers were found that there is no significant influence on the business performance of SMEs in Coimbatore district. A major implication for the findings is that the findings of this research will able to give better understanding for entrepreneurs in addressing the factors which will significantly influence the business performance of SMEs. The study of the factors influencing the business performance of SMEs is critical in understanding the business continuity and growth hence help supporting economies development with in a country. Further research can be conducted through a large scale structured data collection and quantitative analysis to test and verify the findings of this study. This research may prove to be beneficial for the policy makers in delineating some measures to strengthen the SMEs sectors which can extremely contribute towards a strong, stable and prosperous condition.

References

1. Abu Bakar, J Mad, C. A. & Abdul Latif, R. (2006) Liberalization and globalization: a case of Naza and a lesson to SME. *Proceedings of Persidangan Kebangsaan IKS 2006*, Kuala Lumpur: Universiti Utara Malaysia.

2. Aris, N. M. (2006) *SMEs: building blocks for economic growth.* [On-line].Available: http://www.statistics.gov.my/conference/download.php?cat=1&id_file=22.
3. Beck, T. & Demirguc-Kunt, A. (2006) Small and Medium Size Enterprise: Access to Finance as a Growth Constraint. *Journal of Banking & Finance*, 30 (1), pp. 2931-2943.
4. Benzing, Chu, H.M. and Kara, O. (2009) Entrepreneurs in Turkey: A factor analysis of motivation success factors and problems. *Journal of Small Business Management*, Vol.47, No.1, pp.58-91.
5. Bird, B. (1988) Implementing entrepreneurial ideas: the case for intention. *Academy of Management Review*, Vol. 13 No. 3, pp. 442-453.
6. Cacciotti, G. and Hayton, J.C. (2015) Fear and Entrepreneurship: A Review and Research Agenda. *International Journal of Management Reviews*, 17 (2), pp. 165-190.
7. Dalberg. (2011) *Report on Support to SMEs in Developing Countries Through Financial Intermediaries.* Available at: http://www.eib.org/attachments/dalberg_sme-briefing-paper.pdf. Accessed on 21 July 2014.
8. Delmar, F. & Wiklund, J. (2008) The Effect of Small Business Managers' Growth Motivation on Firm Growth: A Longitudinal Study. *Entrepreneurship Theory and Practice*, 32 (3), pp. 437-457.
9. Fatoki, O. & Garwe, D. (2010) Obstacles to the growth of new SMEs in South Africa: a principal component analysis approach. *African Journal of Business Management*, 4 (5), pp. 729-738.
10. Foon, L,S. & Eu-Gene, S. (2006) The changing of education in the globalized world. *Proceedings of Persidangan Kebangsaan IKS 2006*, Kuala Lumpur: Universiti Utara Malaysia.
11. Harvie, C. (2004) *East Asian SME capacity building, competitiveness and market opportunities in a global economy*. [On-line]. Available: http://www.uow.edu.au/commerce/econ/wpapers.html.
12. IC/DIST. (1996) *A Portrait of Australian Business: Results of the1995 Business Longitudinal Survey*, Small Business Research Program, Industry Commission and Department of Industry, Science and Tourism, AGPS, Canberra.
13. Kolstad, I. & Wiig, A. (2015) Education and entrepreneurial success. *Small Business Economics*, 44 (4), pp. 783-796.
14. Ritchie, B. & Brindley, C. (2000) Disintermediation, disintegration and risk in the SME global supply chain. *Management Decision*, 38(8), pp.575-583.
15. Saleh, A.S. & Ndubisi, N.O. (2006) *"SME development in Malaysia: domestic and global challenges"*. [On-line]. Available: http://www.uow.edu.au/commerce/econ/wpapers.html
16. Samad, N. A. (2007) Positioning Malaysian SMEs in the global". *Proceedings of Persidangan Kebangsaan IKS 2007*, Kota Kinabalu: Universiti Utara Malaysia.
17. Scarborough, N.M. Wilson, D.L. & Zimmerer, T.W. (2009) *Effective Small Business Management*. 9th ed. New Jersey: Pearson Education, Inc, Upper Saddle River.
18. SMIDEC (2007) *SME development programme*. [On-line].Available: http://www.smidec.gov.my/detailpage.jsp?section=defsme&level=1.
19. Specht, P.H. (1993) Munificence and carrying capacity of the environment and organization formation. *Entrepreneurship Theory and Practice*, Vol. 17 No. 2, pp. 77-86.
20. Stefanovic, I., Damnjanovic, P. and Jasko,O (2010) The analysis of contemporary Environment impact upon organisational operation. *Serbian Journal of Management*, Vol.5, No.1 pp.97-109.
21. Stuti, K.I. (2005) *Overcoming barriers to innovation for Indian SMEs*. Ministry of Small Scale Industries, New Delhi, India. [On-line]. Available: http://74.125.155.132/scholar?q=cache: ZyqqVy8Bof4J:scholar.google.com/+Stuti+%2B+Overcoming+barriers+to+innovation+for+Indian+SMEs& hl=en.
22. Van Eeden, S., Viviers, S. & Venter, D. (2003) A comparative study of selected problems encountered by small businesses in the Nelson Mandela, Cape Town and Egoli Metropoles. *Management Dynamics*, 12 (3), pp. 13-23.
23. Venter, J.M.P. & de Clercq, B. (2007) Tax compliance burden for small, medium and micro establishments in the business services industry. *South African Journal of Economic and Management Sciences*, 10 (1), pp. 1-13.
24. Wafa.S. A., Noordin. R. & Kim-Man, M. (2005) Strategy and performance of small and medium-size enterprises in Malaysia. *Proceedings of the International Conference in Economics and Finance (ICEF)*, Labuan: Universiti Malaysia Sabah.
25. Wang, S. D. (2003) The implications of e-financing: implications for SMEs. *Bulletin on Asia-Pacific Perspective 2003/2004*. New York: United Nations.
26. World Bank, (2006) *World Development Indicator Database.* Available at: http://worldbank.org/WBSITE/EXTERNAL/DATASTATISTICS. Accessed on 15 July 2014.
27. Zhang, S.S., van Doorn, J. & Leeflang, P.S. (2014) Does the importance of value, brand and relationship equity for customer loyalty differ between Eastern and Western cultures?. *International business review*, 23 (1), pp. 284-292.

Development of an Entrepreneurial Intention Model using Social Cognitive Career Theory: A Conceptual Framework.

Pappusamy Udayanan

Majan University College, Muscat, Oman

Abstract: Entrepreneurs and entrepreneurship contributes to employment generation, economic value addition, knowledge sharing and skill building. Oman government has various initiatives and support mechanisms to develop entrepreneurship. The government is encouraging young generation, especially the students (young graduates) to become entreprepreneurs. This leads to an interesting research question "Is entrepreneurship an attractive career choice for graduate students?" This study theoretically addresses the above research question using Social cognitive career theory (SCCT) by Lent et al. (1994) focusing on the entrepreneurial intention of graduate students. This theoretical paper is a review of the existing literature to study the role of SCCT on entrepreneurial intention. This paper concludes with an entrepreneurial intention model, focusing of self-efficacy, entrepreneurial self-efficacy based on SCCT, which could be empirically tested and validated.

Key words: *Social Cognitive Career Theory, Entrepreneurial intention, Self-efficacy; Entrepreneurial self- efficacy*

1. Introduction

Oman economy is facing challenging times due to decline in the contribution from the hydrocarbon sector. The export contribution from the hydrocarbon sector reduced to 57.9 percentage in the year 2016 and its contribution to Oman GDP at market prices dropped to 27.4 percentage during the same year (CBO, 2017). Oman's growth was affected by low oil production, low oil prices and weak investment. Oman's participation in oil production cuts in 2017 further reduced the contribution to GDP by 2.8 percentage. Moreover, non-hydrocarbon GDP is expected to slow down to 2.5 percentage in 2017 from 3.4 percentage in 2016 (World Bank, 2017). Public authorities are constantly monitoring and implementing appropriate measures to reduce the impact of low oil prices on the country's economy. The current situation requires huge economic diversification to focus on the non-hydrocarbon sector's growth to create more employment opportunities for the nationals. The employment of Omanis in the public sector is on the downtrend and there is a mounting pressure on the private sector to absorb new graduates and job seekers into the employment system. Employment opportunities ought to be created for young graduates of the country. Government is on full throttle to promote Small and Medium Enterprises as an economic engine, as it would facilitate diversification, promote entrepreneurship and increase employment opportunities. Thurik and Wennekers (2004) asserted that entrepreneurship drives competitive employment creation and promotes economic growth. The entrepreneurial activity of a country contributes to its economic growth (Hafer, 2013). Moreover, not including entrepreneurship as a strategic growth plan in a five-year development plan of a country would eventually lead to unemployment issues (Rasem and Kabir, 2011). Entrepreneurship is a viable alternate for employment generation (Marco et al., 2008). Practitioners and academicians have put in concerted efforts to develop entrepreneurial culture, focusing on graduate entrepreneurship and entrepreneurial intention (Nabi and Holden, 2008 and Krueger et al., 2000). Developing young graduates as entrepreneurs will augment the employment measures adopted by the country through creation of self-employment opportunities. Graduate entrepreneurship is "interaction between the graduate as the product of university education and business start-up in terms of an individual's career-orientation and mind set towards self-employment" (as cited in Nabi and Holden, 2008, pp546-47). Given this context, the most challenging task is to develop graduates who would pursue entrepreneurship as a career. Students from developing economies are more likely to pursue entrepreneurship as their career, when compared to industrialized economies (Davey et al., 2011). The aim of this research review is to provide a theoretical understanding of graduates' entrepreneurship intention using Social Cognitive Career Theory (SCCT) by Lent et al. (1994). The SCCT is widely accepted framework to study entrepreneurial intention (EI) and it is less researched in the Gulf region. This review paper contributes to the existing SCCT framework by adding new constructs, thus broadening the scope of existing framework and provide deeper insight in to determining the entrepreneurial intentions of the graduates. Entrepreneurial intention is based on the premises that stronger the entrepreneurial intention, more likely the entrepreneurial behaviour is exhibited. This paper attempts to primarily understand Social Cognitive Carer Theory (SCCT) through review of existing literature and then explore the role of SCCT in entrepreneurial intention. By considering the role of SCCT in EI, the paper stablished a hypothesised Entrepreneurial Intention (EI) model, which could be empirically tested and validated.

2. The predictor of Entrepreneurialism: Social Cognitive Career Theory

This paper focuses on the cognitive individual related factors of the SSCT i.e., the self-efficacy, outcome expectations, and personal goals. (Lent et al., 2002). Lent et al. (1994) framework/model focused on career interest and its relation to career choices. Self-efficacy, outcome expectations, and personal goals controls one's own career (Lent et al., 2000). SCCT states that self-efficacy influences personal goals, which determines the individual's action based on the expected outcomes. Self-efficacy is one's own beliefs on the capabilities to complete a task successfully (Bandura, 1986, cited in Boyd & Vozikis, 1994). People are discouraged by failures, when they experience easy success. Effort and perseverance is required to overcome the obstacles and to gain stable sense of self-efficacy (Wood & Bandura, 1989). Perseverance is achieving the goals in challenging times. The level of self-efficacy influences the outcome expectations and its success and failures. Self-efficacy determines the level of pessimism and optimism in a person (Bandura, 2012). An individual with high self-efficacy will persevere in challenging times (Bandura, 2002). The thinking and behaviour of individuals with high self-efficacy and low self-efficacy differs (Bandura, 1986, cited in Boyd & Vozikis, 1994). Outcome expectations are the "belief about the consequences of performing a behaviour" (Lent et al., 2002, p.262). Self-efficacy and outcome expectations influence the career choice, goals and ambition. A goal is the conviction to perform an activity that affects an outcome (Bandura, 1986). Individuals with high level of self-efficacy and wish to achieve their expected outcomes will achieve the desired goals. Self-efficacy influences outcome expectations of the students (Lanero et al., 2015). One's choice of career depends on self-efficacy and expected outcomes (Lent et al., 1994). Achieving desired goals creates a positive change, and it positively influences entrepreneurial commitment (Jansen, 2004), entrepreneurial intention (Dinis Anabela et al., 2013) and the need for achievement and persistence (Wu and Dagher, 2007). Successful entrepreneurs are characterised as goal-oriented individuals, during challenging times (Osborne, 1995). Wu and Li (2011) focused on different constructs, namely, the perceived benefits and perceived sacrifices to study entrepreneurial intention that is synonymous with expected outcomes. They found that the perceived benefits of entrepreneurship explained the formation of entrepreneurial intention. Outcome expectations and entrepreneurial intention vary according to an individual's psychological and socio-economic conditions. Entrepreneurial intention is determined by financial security (Marco et al., (2008), need for achievement (Dinis Anabela et al., 2013) and perceived sacrifices (Wu and Li, 2011). Various research perspectives argue that SCCT is a valid framework to study the relationship between self-efficacy and EI (Austin and Nauta, 2016; Liguori et al., 2017). Entrepreneurial intention is an important construct in explaining entrepreneurship development. Many researches on entrepreneurial intention have used Ajzen's (1991) and Ajzen & Fishbein's (2000) theory of planned behaviour (Krueger et al., 2000; Marco et al., 2008; Engle et al., 2010; do PaÃ§o et al., 2011; Zhang et al.,2014; Aloulou,2016; Gorgievski,2017). Intentions influence the choice of actions and predicts future behaviours. Shook et al. (2003) defined the intention of starting a new business as the conscious effort and focused attention of the mind. Buttar (2015) asserted that entrepreneurship as career is a conscious decision and it is based on individual's intention. Many researches reveal relationship between self-efficacy and entrepreneurial intention (Byod & Vozikis, 1994; Segal et al., 2002; Kristiansen and Indarti, 2004; BarNir et al., 2011; Bayrón, 2013; Lanero et al., 2015). In this context, the following propositions are assumed:

P1: A person's self-efficacy influences his/her goals and the outcome expectations;

P2: A person's outcome expectations determines the goals; and

P3: A person's self-efficacy and goals determines entrepreneurial intention.

3. Is self-efficacy an antecedent to entrepreneurial self-efficacy?

Self-efficacy is one's own beliefs to complete tasks successfully. General self-efficacy is useful when an individual faces challenging situations and when wants to set up a firm (Locke & Baum, 2007). Self-efficacy influences a person's behaviour to start a business and the person's level of 'task persistence' during challenging times (Bandura, 1986). Self-efficacy influences a person's' career choices (Lent, et al., 1994). Self-efficacy better explains entrepreneurial intentions and behaviour (Zhao et al., 2005; Wilson et al., 2007; Naktiyok, 2009; Pittaway et al.,2010; Laguna, 2013; Pihie, 2013; Sušanj and Miljković, 2015; Fietze and Boyd, 2017; Gorgievski, 2017). Sequeira et al. (2007) also found that high self-efficacy leads to increased entrepreneurial intention and the willingness to start business. On the other hand, students ranking high on self-efficacy and pursuing practical course showed more inclination towards entrepreneurship than students ranking high on self-efficacy and pursuing theoretical courses (Piperopoulos and Dimov, 2015). Individuals with higher self-efficacy in their early stages of career will possess high entrepreneurial intentions and will be involved in entrepreneurial tasks (Boyd and Vozikis 1994). Self-efficacy could be general or domain/task-specific. Domain/task-specific self-efficacy is a person's ability to perform the various entrepreneurial tasks effectively (Chen et al., 1998). Entrepreneurial self-efficacy (ESE) is one's own ability to start and successfully manage the venture with required entrepreneurial skills in marketing, human resources, finance, etc. (Chen et al., 1998). ESE predicts entrepreneurial interests and intentions (Chen et al., 1998; Krueger, et al., 2000; Zhao et al., 2005; Wilson et al., 2007; BarNir, et al., 2011; Phie and Badheri, 2013) and predicts entrepreneurial activity (Boyd and Vozikis, 1994). McGee (2009) stated that it is important to include ESE in the EI model. Numerous studies support that 'person with high entrepreneurial self-efficacy has high entrepreneurial intention'. Krueger and Brazeal (1994) developed an entrepreneurial intention model, where perceived venture self-efficacy was included that is synonymous with the perceived entrepreneurial self-efficacy. Various studies argue that self-efficacy and perceived entrepreneurial self-efficacy significantly contribute towards entrepreneurial intention. Nevertheless, the question that arises here is "Is self-efficacy an antecedent to perceived entrepreneurial self-efficacy?" Based on this, the following propositions are assumed:

P4: Self-efficacy is an antecedent to entrepreneurial self-efficacy.

P5: Entrepreneurial self-efficacy of a person influences his/her entrepreneurial intention.

4. The role of Gender

The role of women in socio-economic development is inevitable. Women empowerment and gender equality is prominent in Oman (Ghouse et al, 2017). Women constitute 34.16 percentage of the total population with 24 years as female median age (WPR, 2018). A survey by the Ministry of Higher Education found that women are less likely to find an employment Times of Oman, 2015). Hence, it becomes imperative to study the role of gender in entrepreneurial intentions. Many gender-based researches on entrepreneurial self-efficacy and entrepreneurial intention has been carried out. Women have low self-efficacy than men (Gatewood et al., 2002; Wilson et al., 2007). Women have scored low on the entrepreneurial self-efficacy, entrepreneurial intentions, and entrepreneurial career intention (Chen et al., 1998; Zhao et al., 2005; Wilson et al., 2007;Haus et al., 2013; Kurczewska, 2014;). Men show higher ESE than women (Wilson et al., 2007; Kelly et al., 2012; Arshad et al., 2016). Hence, women may not exploit the entrepreneurial opportunities (Kelly et al., 2012) and are less inclined towards entrepreneurial activity (Canizares and Gracia, 2010). Relationship was also evident between ESE and EI among women students in a variety of majors (Shinnar et al., 2014Austin and Nauta, 2016). However, significant differences could not be established in perceived self-efficacy beliefs between male and female students (Mueller and Conway Dato-on, 2008; Kurczewska and Bialek, 2014). These results provide a strong evidence to develop a proposition on the role of gender on self-efficacy, perceived entrepreneurial self-efficacy and entrepreneurial intentions.

P5: Self-efficacy, entrepreneurial self-efficacy and entrepreneurial intention are gender dependent.

5. EI model, its implication and future research: A conclusive remark.

Based on the various propositions, the proposed EI model is shown in figure1.

Figure 1: Entrepreneurial Intention research model

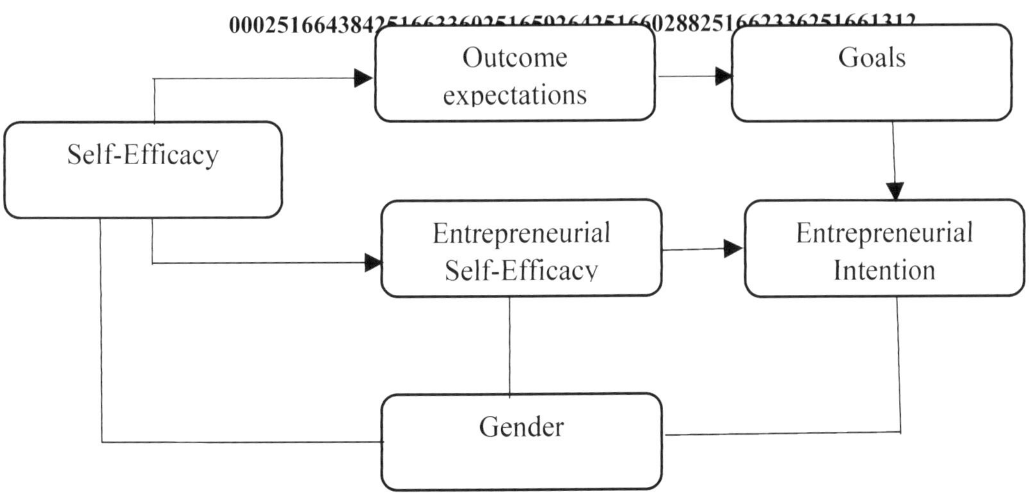

In light of this model, using cognitive individual related factors of the SCCT, including ESE as a precedent to self-efficacy, and incorporating the role of gender in the expanded model, will definitely contribute to the existing knowledge in entrepreneurship research. Testing this model with young graduates will have implications for entrepreneurship development. The findings will contribute to the field of entrepreneurship education focusing of gender, and it could provide valuable insights on the role of self-efficacy and entrepreneurial self-efficacy based on which the role of psychological constructs can be thoroughly investigated for greater understanding. As future development, the role of situational factors like environment, education, infrastructural support, resource capabilities and socio-cultural factors could be included in this model to develop a comprehensive framework to research entrepreneurship.

References:

Ajzen, I. & Fishbein, M. (2000) 'Attitudes and the Attitude-Behavior Relation: Reasoned and Automatic Processes', European Review of Social Psychology, vol. 11, no. 1, pp. 1-33.

Ajzen, I. (1991) 'The theory of planned behavior', Organizational Behavior and Human Decision Processes, vol.50, no.2, pp 179– 211.

Aloulou, (W.J. 2015) 'Entrepreneurial Intention among Freshmen Students - Application of the Theory of Planned Behaviour in Saudi Context', Journal of Enterprising Culture, vol. 23, no. 4, pp. 473-500.https://doi.org/10.1108/ JSBED-02-2016-0028

Arshad, M., Farooq, O., Sultana, N. & Farooq, M. (2016) 'Determinants of individuals' entrepreneurial intentions: a gender-comparative study', Career Development Internationl, vol. 21, no. 4, pp. 318-339.

Austin, M.J. & Nauta, M.M. (2016) Entrepreneurial Role-Model Exposure, Self-Efficacy, and Women's Entrepreneurial Intentions', Journal of Career Development, vol. 43, no. 3, pp. 260-272.

Bandura, A. (1986) 'Social Foundations of Thought and Action: A Social Cognitive Theory'. Englewood Cliffs, NJ: Prentice-Hall In Boyd, N, & Vozikis, G 1994, ' The Influence of Self-Efficacy on the Development of Entrepreneurial Intentions and Actions', Entrepreneurship: Theory & Practice, 18, 4, pp. 63-77, Business Source Premier, EBSCOhost, viewed 1 November 2017.

Bandura, A. (2002) 'Social cognitive theory in cultural context. Journal of Applied Psychology: An International Review', vol.51, pp. 269–290.

Bandura, A. (2012) 'On the Functional Properties of Perceived Self-Efficacy Revisited', Journal of Management, January, Business Source Premier, EBSCOhost, viewed 1 November 2017.

BarNir, A., Watson, W.E. & Hutchins, H.M. (2011) 'Mediation and Moderated Mediation in the Relationship Among Role Models, Self-Efficacy, Entrepreneurial Career Intention, and Gender', Journal of Applied Social Psychology, vol. 41, no. 2, pp. 270-297. doi:10.1111/j.1559-1816.2010.00713.x

Bayrón, C, E. (2013) 'Social Cognitive Theory, Entrepreneurial Self-Efficacy and Entrepreneurial Intentions: Tools to Maximize the Effectiveness of Formal Entrepreneurship Education and Address the Decline in Entrepreneurial Activity', Revista Griot, Vol.6, no. 1, pp. 66-77

Boyd, N, & Vozikis, G. (1994) 'The Influence of Self-Efficacy on the Development of Entrepreneurial Intentions and Actions', Entrepreneurship: Theory & Practice, vol.18, no.4, pp. 63-77, Business Source Premier, EBSCOhost, viewed 1 November 2017.

Buttar, H. M. (2015) ' Formation of Entrepreneurial Career Intentions: The Role of Socio cognitive Factors', Journal of Employment Counseling (Retreacted), vol. 53, no. 3, pp. 144-144.

Canizares, M.S., & Garcia, F. (2010) 'Gender differences in entrepreneurial attitudes', Equity, Diversity and Inclusion: An International Journal, vol. 29, no. 8, pp. 766-786.

Cbo-oman.org. (2017). [Online] Available at: http://www.cbo-oman.org/annual/AnnualReport2016English.pdf [Accessed 28 Dec. 2017]

Chen, C. C., Greene, P. C., & Crick, A. (1998) 'Does entrepreneurial self-efficacy distinguish entrepreneurs from managers? ' Journal of Business Venturing, vol.13, pp.257–332. doi: 10.1016/S0883-9026(97)00029-3

Davey, T., Plewa, C. & Struwig, M. (2011) 'Entrepreneurship perceptions and career intentions of international students', Education + Training, vol. 53, no. 5, pp. 335-352.

Dinis, A., Arminda, d.P., João Ferreira, Mário Raposo & Ricardo, G.R. (2013) 'Psychological characteristics and entrepreneurial intentions among secondary students', Education + Training, vol. 55, no. 8, pp. 763-780.

do PaÃ§o, Arminda M. Finisterra, Ferreira, J.M., Raposo, M., Rodrigues, R.G. & Dinis, A. (2011) 'Behaviours and entrepreneurial intention: Empirical findings about secondary students', Journal of International Entrepreneurship, vol. 9, no. 1, pp. 20-38.

Engle, R.L., Dimitriadi, N., Gavidia, J.V., Schlaegel, C., Delanoe, S., Alvarado, I., He, X., Buame, S. & Wolff, B. (2010) 'Entrepreneurial intent: A twelve□country evaluation of Ajzen's model of planned behavior', International Journal of Entrepreneurial Behavior & Research, Vol. 16 Issue: 1, pp.35-57, https://doi.org/10.1108/13552551011020063

Fietze, S., & Boyd, B. (2017) 'Entrepreneurial intention of Danish students: a correspondence analysis', International Journal of Entrepreneurial Behavior & Research, Vol. 23 no. 4, pp.656-672, https://doi.org/10.1108/IJEBR-08-2016-0241

Gatewood, E., Shaver, K., Powers, J., & Gartner, W. (2002) 'Entrepreneurial expectancy, task, effort and performance', Entrepreneurship Theory and Practice, vol.27 (winter), pp. 187–206.

Ghouse, S., McElwee, G., Meaton, J. & Durrah, O. 2017, 'Barriers to rural women entrepreneurs in Oman', International Journal of Entrepreneurial Behavior & Research, vol. 23, no. 6, pp. 998-1016.

Gorgievski, M.J., Stephan, U., Laguna, M. & Juan, A.M. (2017) 'Predicting Entrepreneurial Career Intentions: Values and the Theory of Planned Behavior', Journal of Career Assessment, pp. 1-19 DOI: 10.1177/1069072717714541

Hafer,R.W. (2013) 'Entrepreneurship and state economic growth', Journal of Ent and Pub Pol, vol. 2, no. 1, pp. 67-79.

Haus, I., Steinmetz, H., Isidor, R. & Rüdiger Kabst (2013) 'Gender effects on entrepreneurial intention: a meta□analytical structural equation model', International Journal of Gender and Entrepreneurship, vol. 5, no. 2, pp. 130-156.

Jansen, K. (2004) ''From persistence to pursuit: a longitudinal examination of momentum during the early stages of strategic change'', Organization Science, Vol. 15 No. 3, pp. 276-94

Kristiansen, S. & Indarti, N. (2004) 'Entrepreneurial intention among Indonesian and Norwegian students', Journal of Enterprising Culture, vol. 12, no. 01, pp. 55-78.

Krueger, N.F. & Brazeal, D.V., (1994) 'Entrepreneurial potential and potential entrepreneurs', Entrepreneurship theory and practice, vol.18, pp.91-91.

Krueger, N.F., Reilly, M.D. & Carsrud, A.L. (2000) 'Competing models of entrepreneurial intentions', Journal of Business Venturing vol.5/6, pp. 411–432.

Kurczewska, A. & Bialek,J. (2014) 'Is the interplay between self-efficacy and entrepreneurial intention gender-dependent?', Argumenta Oeconomica, vol. 2, no. 33, pp. 23-38.

Laguna, M. (2013) 'Self-efficacy, self-esteem, and entrepreneurship among the unemployed', Journal of Applied Social Psychology, vol. 43, no. 2, pp. 253-262.

Lanero, A., Jose-Luis Va¡zquez & Carlota, L.A. (2016) 'Social cognitive determinants of entrepreneurial career choice in university students', International Small Business Journal, vol. 34, no. 8, pp. 1053-1075.

Lent, R. W., Brown, S. D., & Hackett, G. (1994) 'Toward a unifying social cognitive theory of career and academic interest, choice, and performance', Journal of Vocational Behavior, 45, 79–122. doi:10.1006/jvbe.1994.1027

Lent, R.W., Brown, S.D. & & Hackett, G. (2002) 'Social Cognitive Career Theory' in Career Choice and Development, 4th edn, Duane Brown and Associates, Wiley, SF-CA.

Lent, R.W, Brown, S.D. & Hackett, G. (2000) 'Contextual Supports and Barriers to Career Choice: A Social Cognitive Analysis', Journal of Counselling Psychology, Vol. 47, No. 1, pp. 36-49

Liguori, E., Bendickson, J., S. & McDowell, W. (2017) 'Revisiting entrepreneurial intentions: a social cognitive career theory approach Revisiting entrepreneurial intentions: a social cognitive career theory approach', pp.1-12 International Entrepreneurship and Management Journal, https://doi.org/10.1007/s11365-017-0462-7

Locke, E. A., & Baum, J. R. (2007) 'Entrepreneurial motivation'. In Laguna, M. 2013, 'Self-efficacy, self-esteem, and entrepreneurship among the unemployed', Journal of Applied Social Psychology, vol. 43, no. 2, pp. 253-262.

Marco, v.G., Brand, M., Mirjam, v.P., Bodewes, W., Poutsma, E. & Anita, v.G. (2008) 'Explaining entrepreneurial intentions by means of the theory of planned behaviour', Career Development International, vol. 13, no. 6, pp. 538-559.

McGee, J.E., Peterson, M., Mueller, S.L. & Sequeira, J.M. (2009) 'Entrepreneurial Self-Efficacy: Refining the Measure', Entrepreneurship Theory and Practice, vol. 33, no. 4, pp. 965-988.

Nabi, G. & Holden, R. (2008) 'Graduate entrepreneurship: intentions, education and training', Education and Training, Vol. 50 No. 7, pp. 545-51.

Naktiyok, A., Karabey, C. & Gulluce, A. (2009) 'Entrepreneurial self-efficacy and entrepreneurial intention: The Turkish case', International Entrepreneurship and Management Journal, Vol 6, no. 4, pp 419–435

Times of Oman (2015) 'MoHE survey shows employment rate low among women graduates – Oman' [Online] Muscat Daily News. Available at: http://www.muscatdaily.com/Archive/Oman/MoHE-survey-shows-employment-rate-low-among-women-graduates-410k [Accessed 2 Jan. 2018].
Osborne, R.L. (1995) 'The essence of entrepreneurial success', Management Decision, Vol. 33, pp. 4-9.

Pihie, Z.A.L. & Bagheri, A. (2013) 'Self-Efficacy and Entrepreneurial Intention: The Mediation Effect of Self-Regulation', Vocations and Learning, vol. 6, no. 3, pp. 385-401.

Piperopoulos, P & Dimov, D. (2015) 'Burst Bubbles or Build Steam? Entrepreneurship Education, Entrepreneurial Self-Efficacy, and Entrepreneurial Intentions', Journal of Small Business Management vol.53, no.4, pp. 970–985

Pittaway, L., Elena Rodriguez-Falcon, Aiyegbayo, O. & King, A. (2011) 'The role of entrepreneurship clubs and societies in entrepreneurial learning', International Small Business Journal, vol. 29, no. 1, pp. 37-57.

Rasem, N.K. & Kabir Hassan, M. (2011) 'Saudi Arabia's economic development: entrepreneurship as a strategy', I J Islam Mid East Fin and Mgt, vol. 4, no. 1, pp. 52-73.

Segal, G, Borgia, D, & Schoenfeld, J. (2002) 'Using Social Cognitive Career Theory to Predict Self-Employment Goals', New England Journal of Entrepreneurship: Vol. 5: No. 2. Article.8 (Available at: http://digitalcommons.sacredheart.edu/neje/vol5/iss2/8)

Sequeira, J., Mueller, S.L. & Mcgee, J.E. (2007) 'The Influence of Social Ties and Self-Efficacy in Forming Entrepreneurial Intentions and Motivating Nascent Behavior', Journal of Developmental Entrepreneurship, vol. 12, no. 3, pp. 275-293.

Shinnar, R.S., Hsu, D.K. & Powell, B.C. (2014) 'Self-efficacy, entrepreneurial intentions, and gender: Assessing the impact of entrepreneurship education longitudinally'. The International Journal of Management Education, Vol.12, no. 3, pp. 561-570

Shook, C.L., Priem, R.L. and McGee, J.E. (2003) 'Venture creation and the enterprising individual: A review and synthesis', Journal of management, vol.29, no.3, and pp.379-399.

Sušanj, Z., Jakopec, A. & Miljković Krečar, I. (2015) 'Verifying the model of predicting entrepreneurial intention among students of business and non-business orientation', Journal of Contemporary Management Issues, Vol. 20, pp.49-69.

Thurik, R. & Wennekers, S. (2004) 'Entrepreneurship, small business and economic growth', Jrnl of Small Bus Ente Dev, vol. 11, no. 1, pp. 140-149.

Wilson, F., Kickul, J. & Marlino, D. (2007) 'Gender, Entrepreneurial Self-Efficacy, and Entrepreneurial Career Intentions: Implications for Entrepreneurship Education1', Entrepreneurship Theory and Practice, vol. 31, no. 3, pp. 387-406.

Wood, R. & Bandura, A. (1989) 'Social cognitive theory of organizational management', Academy of management Review, vol. 14, no. 3, pp. 361-384.

World Bank. (2017). Oman's Economic Outlook - October 2017. [Online] Available at: http://www.worldbank.org/en/country/gcc/publication/oman-economic-outlook-october-2017 [Accessed 28 Dec. 2017]

Worldpopulationreview-WPR.(2018). [Online] Available at: http://worldpopulationreview.com/countries/oman-population/ [Accessed 2 Jan. 2018]

Wu, L. & Li, J. (2011) 'Perceived value of entrepreneurship: A study of the cognitive process of entrepreneurial career decision', Journal of Chinese Entrepreneurship, vol. 3, no. 2, pp. 134-146.

Wu, S., Matthews & Dagher, G.K. (2007) 'Need for achievement, business goals, and entrepreneurial persistence', Management Research News, vol. 30, no. 12, pp. 928-941.

Zhang, P., Wang, D. & Owen, C. (2014) 'A Study of Entrepreneurial Intention of University Students', Entrepreneurship Research Journal, vol.5, no.1, pp. 61-82. Retrieved 1 Jan. 2018, from doi: 10.1515/erj-2014-0004.

Zhao, H., Seibert, S. E., & Hills, G. E. (2005) 'The Mediating Role of Self-Efficacy in the Development of Entrepreneurial Intentions', Journal of Applied Psychology, *vol.90*, no.6, pp.1265-1272. http://dx.doi.org/10.1037/0021-9010.90.6.1265

PROTECT OUR COMPUTERS AGAINST CYBERCRIME

Safa Khalfan Al-Dhafri, Khalid Juma Al-Mur Al-Zidi
B.Tech,Student,Dept.IT,Ibra College of Technology,Oman
36j131@ict.edu.om , 32J12122@ict.edu.om

Abstract

Cybercrime is a complex and ever-changing incident. Moreover, it is becoming more serious. Finding from 2002. Cybercrimes it can be defined as: illegal act that are funds against individuals or groups of individuals with a criminal motive to intentionally harm the reputation of the victim or cause physical or mental harm, or loss, to the victim directly or indirectly, using modern telecommunication networks such as Internet (networks including but not limited to Chat rooms, emails, notice boards and groups) and mobile phones (Bluetooth/SMS/MMS)[6]. This paper analyse the understanding of Cybercrime of 50 person from my college who are (students from different levels and staffs).My objective is to find out more are understand what is cybercrime and how to protect themselves against it.

Introduction

Smart phones, personal computers, tablets and mobile device etc. come under cyberspace. Cyberspace is electronic medium of computer network in which on-line communication take place. Every moment we begin to modify our lives by using the new technologies. In the course of time, we may knowingly or unknowingly will face some kind of trap. The most common one of these traps is cybercrime. The term of cyber is derived from "cybernetic" that comes from Greek ,which mean "skilled in governing" . It is widely used by many information technology industries. So now it has become a recent term in internet area. Secondly, the term 'crime' means illegal act by a state or other authority. Cybercrime use computer and network to steal person's identities. The category of cybercrime can be categorized in two way: first a computer as a target, by using a computer to attack other computers for example hacking ,virus/worm ,denial of service attack and Trojan horse attack etc. Second way is use a computer as a weapon, by using a computer to commit real world crime for example cyber terrorism, credit card fraud and pornography. According to reliable websites and books the first cybercrime was recorded in 1978. "The technology is like a fish, the longer it stay on the shelf, the less desirable it become "by Alan Watts.

Literature Review

TOP 5 CYBERCRIME

- Tax-Refund Fraud
- Corporate Account Takeover
- Identity theft.
- Theft of sensitive data.
- Theft of intellectual property..

A Study of Cyber Security Challenges and Its Emerging Trends on latest technology

- IPv6: It is a new internet protocol, which is replacing IP4 that is old version.
- more users in social media.
- many cyber ethics guide users in using the internet in a proper and safer way

A Survey of Cyber Crime in Greece

- Survey of cybercrimes in Greece is discussed. Now day's cyber space is in crowded because of social media, e-commerce- , e banking etc.

Cyber Crime: Prevention & Detection

- We should know how to detect and prevent cybercrime, Then we are in a position to defend against cybercrime.

Organizations and Cyber crime.

- This group strives to accomplish a different kind of goals, including defiance of authority and technology.

Awareness about Cybercrime at Student Level and others

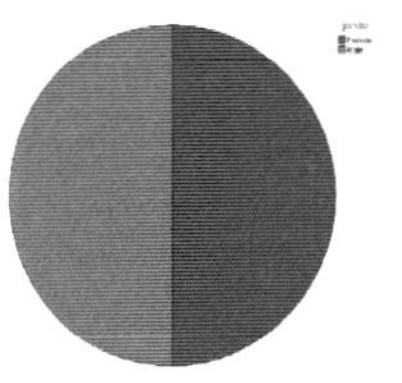

Figure(1) Percentage of Female & Male

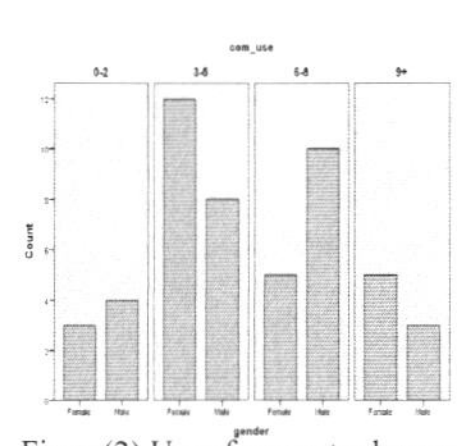

Figure(2) Use of computer hours per day.

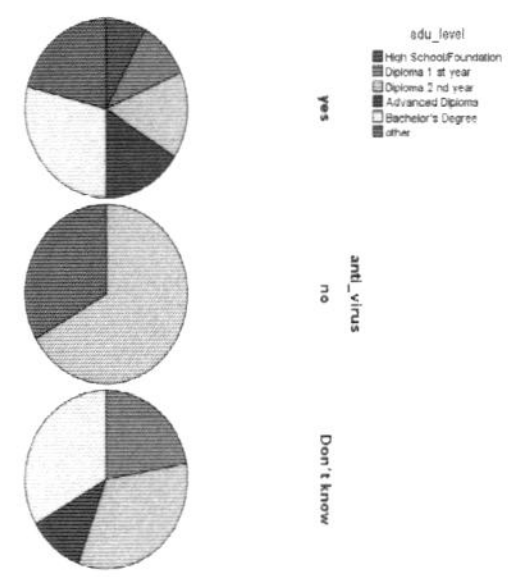

Figure(3) Installation of the anti-virus program

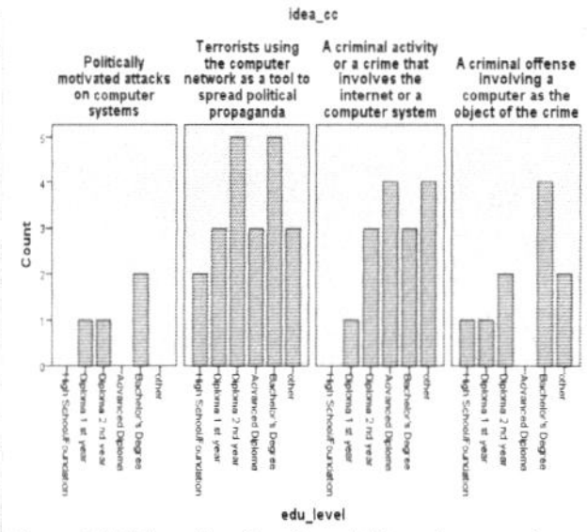

Figure(4)Education level and idea about cybercrime.

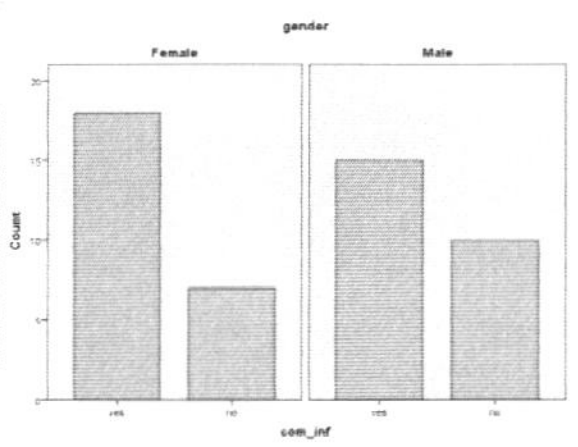

Figure(5)Users infected by cybercrime

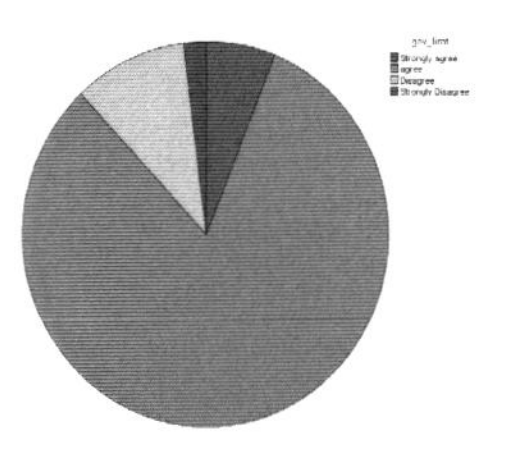

Figure(6)Increased Internet surveillance

Cybercrime

1.Cyber Security

In cyber security, it very important to be completely with confidentiality, integrity & availability.

2.Social network's Cybercrimes

Cybercriminals also common attack site around our region like many social networking websites.

3.How to avoid Cybercrimes

- Know the falls of public network.
- Use credit cards and do not debit cards.
- 4.4 Avoid junk and unwanted e-mails

Conclusion

- In brief we see from our survey the people knowledge about cybercrime is that not much so I think they have to now deferent e-crime and who can avoid, detect, prevent it . And the people who's have high level of education have more knowledge about cybercrime and how protect them self against cybercrimes.

References:

- [1] Tommie Singleton, "The Top 5 Cybercrimes", American Institute of CPAs. All rights reserved, October 2013, pg 1-15.
- [2] G.Nikhita Reddy , G.J.Ugander Reddy, "A Study of Cyber Security Challenges and Its Emerging Trends on Latest Technologies", International Journal of Engineering and Technology Volume 4 No. 1, January, 2014, ISSN 2049-3444, Pg 48-51.
- [3] Alexandros Papanikolaou, Vasileios Vlachos, Anastasios Papathanasiou, Konstantinos Chaikalis, Maria Dimou, and Magdalini Karadimou," A Survey of Cyber Crime in Greece, Telfor Journal, Vol. 6, No. 2, 2014,pg 86-91.
- [4] Ms M Lakshmi Prasanthi, Tata A S K Ishwarya," Cyber Crime: Prevention & Detection", International Journal of Advanced Research in Computer and Communication Engineering Vol. 4, Issue 3, March 2015, ISSN (Online) 2278-1021 ISSN (Print) 2319-5940 ,pg 45-48.
- [5] Roderic Broadhurst, Peter Grabosky, Mamoun Alazab & Steve Chon," Organizations and Cyber crime: An Analysis of the Nature of Groups engaged in Cyber Crime", International Journal of Cyber Criminology Vol 8 Issue 1 January - June 2014, (IJCC) ISSN: 0974 – 2891,pg 1-2.

An Idea about Smart Cities

Sada Saud Humaid Al-harbi, Advanced Diploma, Dept. of IT, IBRA college of Technology
, 36J1488@ict.edu.om
Amira Hamed Al-Batrani, Advanced Diploma, Dept. of IT, IBRA college of Technology
,36J149@ict.edu.om

Abstract

Internet of Things (IOT), which will create a huge network of "Things" communicating with one another so, Smart Things is one of this Technology shift in IT world. Smart Things are the things that are having embedding smartness such as :smart phone and laptop ,by this things you can identification, automation, monitoring and controlling caliber. Smart Things are making a human life to become easier and fast to do a lot of works on time, in nowadays without applications life is becoming cumbersome and slow. In this world, any new things come especially in Technology world will face many technical and application challenges. So, This paper discussed the definitions of smart cities such as: Muscat, the status of IOT developments and directions of Smart Cities, Characteristics and factors of a smart city, including applications, Who to implement IOT in smart cities, Challenges, problems & benefice which related Future Technology and based on survey from student .Finally, this report discusses how to implement IOT at cities and prospect of IOT.

Introduction

Internet of Things is defined as "An open and comprehensive network of intelligent objects that have the capacity to auto-organize, share information, data and resources, reacting and acting in face situations and changes in the environment". Internet of Things is one of the last advances in Information and Communication Technologies, providing global connectivity and management of sensors, devices, users and information.

Literature Review

Smart Cities & Direction of Smart Cities

- Smart Cities make our world smarter.
- It's consist of a group of devices which can be monitored and controlled via a central processors and web services.
- It is one of the last advances in Information and Communication Technologies.
- providing global connectivity

Related Works

- [1] we discussed about Definitions of Smart Cities.
- [2] we discussed about the Direction of Smart Cities.
- [3] we discussed about properties and factors of a smart city.
- [4] we discussed about Application we covered many applications.
- [5]we researched about Result which is coming under the (IOT) of smart cities .

Characteristics and factors of a smart city

- **Smart Economy**
- **Smart People**
- **Smart Governance**
- **Smart Mobility**
- **Smart Living**
- **Smart Environment**

Application

Internet of things promises a lot of applications in human life, making their life more easier, safe and smart. There are many applications like:

1 Smart cities
2 Smart Home and Buildings
3 Smart Energy and the Smart Grid .

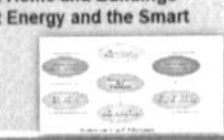

Result

- It operations management can be the lead architects of business transformation.
- To be successful, governments must build the right network infrastructure, imbed context from IOT into business workflows and design a foundation for gathering and understanding sensor data,

Awareness about "An Idea about Smart Cities" at Student Level

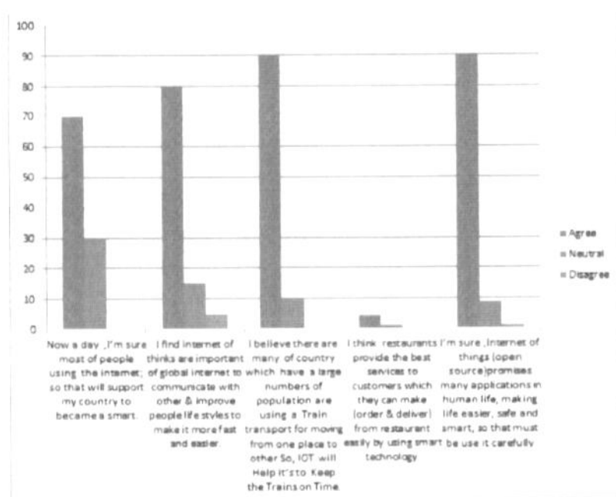

Figure 2 Result Analysis of the An Idea about smart Cities.

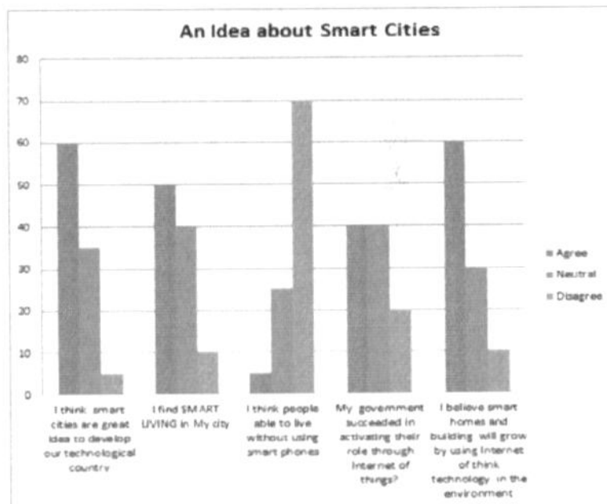

Figure 3 Result Analysis of the An Idea about smart Cities

This chart reviews the results we have obtained through our survey of 100 students from ICT & different department, which includes 10 general questions about smart cities and how the government and its role in activating smart cities to develop the country and individuals.

Challenges , Benefits of "An Idea about Smart Cities"

Wireless communications

You can establish wireless technologies from an energy point of view, such as GSM, UMTS, Wi-Fi and Bluetooth are far less suitable

Security and privacy

A security and protection aspects of the Internet such in communications confidentiality, the authenticity and trustworthiness of communication partners.

Improving Parking & Transportation with IOT

using IOT will help smart cities to manage transportation.

IOT Helps Keep the Trains on Time

use trains to transport people from one place to another and meet the specified time.

Makes order from restaurant

provide the best services to customers, which is the delivery order to homes quickly and high quality in taste one time

Conclusion

- Now a day the world is became as a small city, which is covering the (IOT), that consist of Smart Things in our environments. Human nature prefers exchange in daily routine so, in that case; A new technology coming in internet world and technological development which is discussed about it in this paper. A new technology is including Smart Things that are carrying smart Cities. smart Cities is playing an active role of human life on their everyday. In addition, these applications are countless. In brief we see from our survey The people have a high level of education have more knowledge about "An Idea about Smart Cities" and how to use IOT technology to improve them self.

References:

- [1]https://arxiv.org/ftp/arxiv/papers/1501/1501.02211.pdf
- [2]http://www.spatialcomplexity.info/files/2013/07/BATTY-AESOP-ACSP-20131.pdf
- [3]http://smartcities.eu/download/smart_cities_final_report.pdf
- [4]www.cisco.com/.../industries/docs/gov/everything-for-cities.pdf
- [5]www.ijfcc.org/vol4/395-ICNT2014-2-203.pdf
- [6]https://www.researchgate.net/profile/Elmustafa_Sayed_Ali_Ahmed/publication/313651150_Internet_of_Things_Applications_Challenges_and_Related_Future_Technologies/links/58a6e9b64585150402f2785/Internet-of-Things-Applications-Challenges-and-Related-Future-Technologies.pdf
- [7]https://www.cisco.com/c/dam/en_us/solutions/trends/iot/docs/smart_cities_are_built_on_iot_lopez_research.pdf.

Audit Expectations Gap in the Public Sector of Libya

Tarek Mami[1]

University of Zawia, Libya

Abstract

The purpose of this research is to investigate whether an audit expectations gap exists in the public sector in Libya. The motivation for carrying out this research in Libya was due to the lack of research on this issue. The study concludes with a number of academic and practical recommendations.

Keywords: Auditing, Government agencies, Annual accounts, Public sector organizations

[1] Assistant professor in accounting, faculty of economic, Zawia university, Libya.

Introduction

The report of the Audit Bureau represents the outcome of the audit process from the moment of the issuance of the Budget Law and the Audit Bureau to review until the moment of forming the technical opinion on the books, records, contracts, etc. and to the issuance of the report of the Audit Bureau. Giving information to the reader of the report that the budget for the year as such has been examined and verified by the Audit Bureau. On the other hand, the existence of audit in itself greatly reduces the motivation of making mistakes and manipulating accounting figures by parties within the government sector. The report of the Audit Bureau may be considered as the means of communication between the Bureau and the legislature.

The users of the financial statements are exposed to heavy losses due to fraud or failure, followed by the rise of voices calling the external auditor. Most of the blame and accusation is directly directed at the external auditor, believing that most of the users of the financial statements, the auditor is the first and last responsible for detecting cases of fraud. The example, in the money-making companies and the Credit and Trade Bank of Egypt, the collapse of Enron Energy and the collapse of Arthur Anderson in America during the early years of the 21st century could be attributed in the prevention and detection of fraud lies primarily in the management through the implementation of the internal control system.

The financial crisis has raised the interest of researchers and politicians to look for the causes of these scandals and collapses in large companies, and one of the most important of these reasons is the so-called "gap expectations The first use of the term “expectations gap” is attributed to Liggio (1974) although the basic idea was in existence before 1974 (Lee,1970). The Cohen Commission (1978, p.1) defined the audit expectations gap as “the gap between what the public expects or needs and what auditor scan and should reasonably expect to accomplish.”

The existence of the expectations gap makes the auditor questioned by the different parties within the community, leaving the auditor with the spirit of initiative and development because he believes that independence and lack of independence exist in an environment that does not guarantee its rights. and perhaps the most important groups benefiting from the profession of government auditing, which has a direct impact on the profession through its own authority, are the legislative, executive and supervisory bodies, and therefore a gap in expectations in these categories will inevitably affect the future of the audit profession in General and governmental ones in particular.

Research question

The main categories benefiting from the profession of government auditing, which has a direct impact on the profession through its own authority, are the legislative, executive and supervisory agencies. Therefore, an expectation gap in these categories will inevitably affect the future of the audit profession in general and the government in particular.

In this study, the legislative body is the financial and planning committee of the state, the executive body represented by the financial departments by the governmental sectors, while the supervisory body represented in the members of the Audit Bureau of the Libyan Audit Bureau. The three parties mentioned above represent the parties involved in this study. The bottom line is that the study problem crystallizes in the following questions:

1. How effective does it take to realize the responsibility of the auditor to narrow the expectations gap?

2. What is the extent to which the Audit Bureau applies independence to both its functional and financial aspects?

3. Are the reports of the Audit Bureau and its emphasis on the principle of transparency and accountability available?

Objectives of the research

The objective of this study is to document and analyze the expectations gap in the government review among three main parties: the planning and financial committees of the legislature and academics, financial managers in the government sectors and members of the Audit Bureau. The gap in the expectations gap between the participating parties will be reviewed in order to identify whether there is a gap. Thus, the objectives of the study can be crystallized in the following points:

1. To recognize the existence of a gap of expectations or not in the responsibility of the auditor and the duties in the examination of all revenues and expenses of the State according to the standard;

2. To identify the appropriateness of the content of the report issued by the Audit Bureau and express it in a transparent manner to meet the requirements of users and beneficiaries and;

3. To identify the extent of cooperation of the government departments with the Audit Bureau and to provide them with all required data.

SIGNIFICANCE OF THE STUDY

This study is important because it is the first locally attempt to study the expectations gap in government auditing in the Libyan environment by addressing the opinions of three important segments of the auditing profession namely: financial managers in the government sectors, members of the Audit Bureau and the financial committees of the state legislature and academics. The importance lies in the following:

1. It contributes to the submission of proposals from the views of the relevant parties and;

2. It will positively contribute to further studies and research in the area of the government expectations gap.

Literature review

There is a need for an empirical research to explore whether or not an audit expectations gap exists in the public sector between auditors and users of their audit reports. In the private sector context, the over whelming conclusion of existing research is that there is an audit expectations gap (Fadzly and Ahmad, 2004; Linand Chen, 2004). Much earlier, the Cohen Commission (1978) concluded that a gap did exist and them main responsibility lay with the auditing profession rather than with the users. The Commission attributed the primary cause of this gap to the failure of the profession to react and evolve quickly enough to keep pace with the changing business environment. Both the Canadian Institute of Chartered Accountants (1988) and Porter (1991, 1993) viewed the audit expectations gap from an expectations-performance point of view. The two main aspects of the gap were identified as a feasibility or reasonableness gap (related to what were regarded as unreasonable expectations by the public) and a performance gap (related to deficient auditing standards and/ or performance). This paper discusses the audit expectations gap in relation to the public sector and then examines the accountability framework and six audit concepts used in this research study followed by sections on the research design, results and conclusions Expectations gap and public sector Research into the audit expectations gap has centred on the private sector and has generally ignored the public sector. As the raison d'etre of audit is similar for both sectors, it is important to research whether or not there is an audit expectations gap in the public sector. Humphrey (1992,p.1) pointed out “the importance of the expectations issue derives from notions of stewardship and public accountability”. These aspects are as relevant in the public sector as in the private sector (BatraandKaur,1993). Public sector auditors need to consider the information requirements of audit report users as much as private sector auditors. Buttery et al. (1993, p.42) suggested that public sector auditors should have there cipients of their reports in mind at all times whilst preparing such documents and Longley (1992) suggested that those who exercise power in society should be accountable forth at power. The issue of exercising government power bring stothe forea much wider recipient group than is normally understood. Grayetal. (1996,p.10) argued that “the accountability framework is the most useful one for analyzing accounting information transmission in general”. This arguments its comfortably with the public sector audit scenario where by society in general could be regarded as responsible.

In the context of the public sector, the audit function has also changed over time. Traditionally, the public sector audit was concerned with regularity, legality and probity of government agencies expenditures. The main considerations were about ensuring that the money allocated by Parliament was spent according to its purposes and that the accounts were properly presented and conformed to the laws and regulations governed by the agencies. However, with public sector reforms, the public sector administration has become more complex. The traditional audit functions have been expanded to include wider monitoring functions over government agencies. The auditors’ task now, is to examine whether programmes implemented by government agencies have been implemented economically, efficiently and effectively. This is widely known as Performance Audit or Value for Money (VFM) audit (Glynn, 1985; Pollitt et al., 1999; INTOSAI, 2004; NAD 2006). The objective is to ensure that the government agencies are accountable not only for the resources they used but also for the effectiveness with which they used those resources. Accordingly, the public sector audit is now concerned with terms such as 'accountability', 'output', 'efficiency', and 'value for money' (Guthrie and Parker, 1999; Ng, 2002).

Research Methodology

The questions in the questionnaire are divided into five groups. The first group of questions is on the demographic profile of respondents. The second group of questions is collected information about The responsibility of government accounts audit. The third group of questions is on the information about report of government accounts audit. The fourth group of questions is on the information about the independent government account audit. The fifth group of questions is the information about Audit evidence.

The second, third, fourth and fifth groups of questions used 5 Likert-scale types of questions.

Kinds of answer	Weight
No strong agree	1
Not agree	2
Somewhat Agree	3
Agree	4
Strong agree	5

Likert scale of 5 points to measure the degree of respondent's agreement were selected.

Measure	arrangement	Standard degree	Percentage(%)	evaluate
Likert fifth	1	From 1 to less than 1.80	20%-36%	Very weak
	2	From 1.80 to less than 2.60	36%-52%	weak
	3	From 2.60 to less than 3.40	52%-68%	Pass
	4	From 3.40 to less than 4.20	68%-84%	Good
	5	More than 4.20	84%- more	High

Design of study

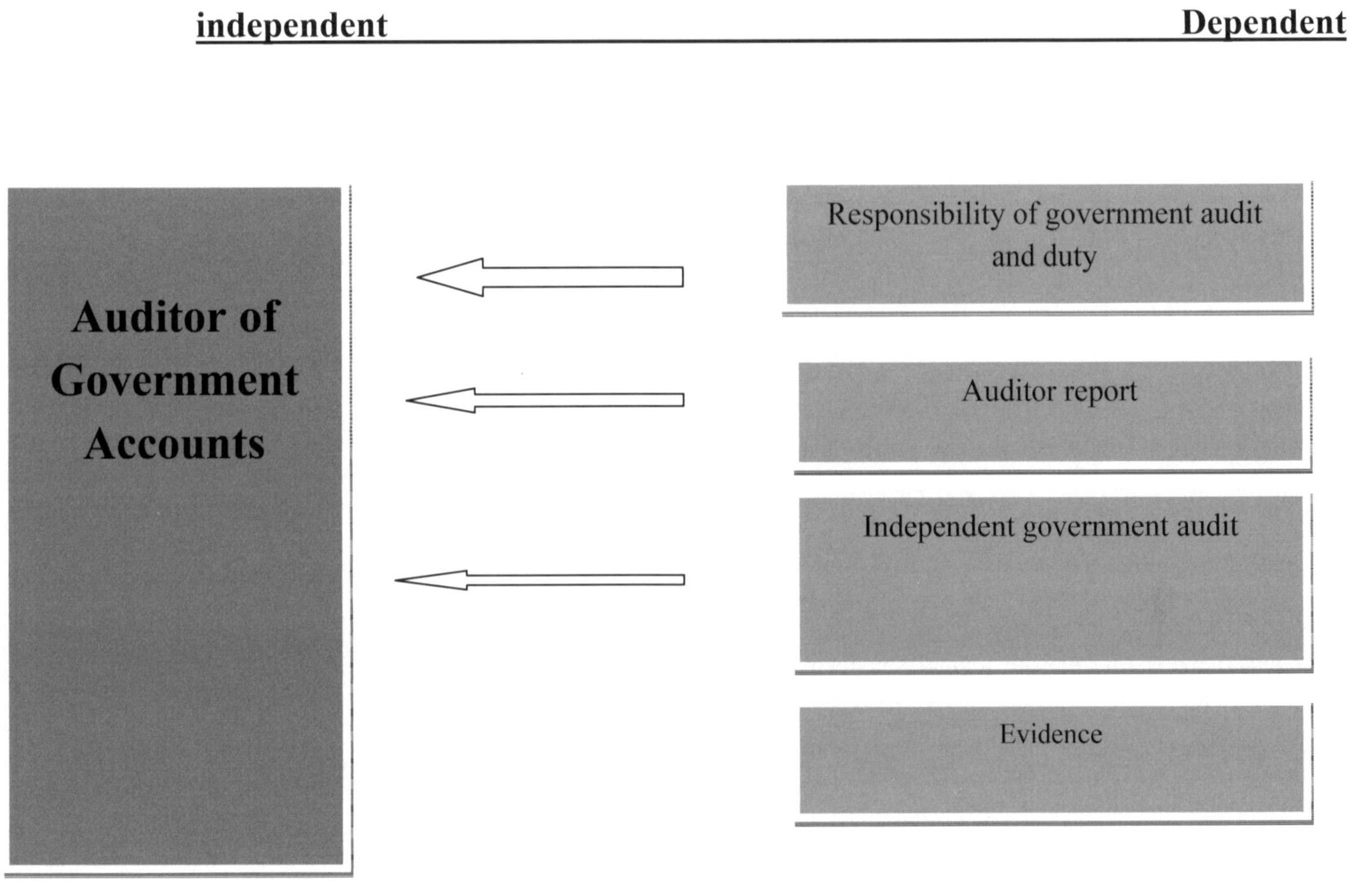

Measurement of variables study

Dependent variable: the government accounts gap expectations in Libya.

Independent variables:

1. First independent variable: The responsibility of government accounts audit by questions from (1-6).

2. Second independent variable: Report of government accounts audit by questions from (7-9).

3. Third independent variable: independent of government account audit by questions from (10-12).

4. Fourth independent variable: Audit evidence by questions from (13-17).

Hypothesis of the study

The basic proposition of the study are the following:

HYPOTHESIS I: There is no difference between the answers from the parts participant at study from the responsibility and duty of government audit.

HYPOTHESIS II: There is no difference between answered from parts participant at study from government audit's report.

HYPOTHESIS III: There is no difference between answers from the parts participant at study from government audit's independent.

HYPOTHESIS IV: There is no difference between answered from parts participant at study from government audit's evidence.

Data collection and sample selection

The survey questionnaire was designed and distributed to get responses from the selected sample study of (i) government audit (ii) financial director of ministries. (iii) Academic and legislative committee. The distributed (170) questionnaires and collected was about (141) questionnaire. Sample size is shown in table no.1.

The selected sample was to try to get perspective of government audit, financial director of ministries, academic and legislative committee.

Table no.1

Items	No. of distributed	No. of responded	No. of unaccounted	Percentage
Government audit	70	61	9	87%
Financial director at ministries	55	45	10	75%
Academic and legislative committee	45	35	10	70%
Total	170	141	29	83%

Reliability test

Table number 2 shows the result of α of all items in the variables is more than 60% that means:

Range	Cronbach's Alpha
Average statistical variables	0.86

The result of Cronbach's Alpha of all items in Government accounts audit gap expectation variable is 0.86. It should be more than 0.6 in the present test as it is more significant with Cronbach's Alpha.

Table Number 3
Demographic variable analysis

Items	Criteria	Frequency	Percentage(%)
Qualification	Master	60	42%
	Bachelor	45	31%
	High diploma	38	27%
Specialization	Financial and administration	134	94%
	Engineering	9	6%
Current position	Head of department	14	10%
	Head of unit	46	32%
	Director	50	35%
	Employment	33	23%
Experience	Less than 5 years	8	6%
	6 to 10 years	11	7%
	More than 10 years	124	87%

Table number 3 significantly projects the qualifications of the staff as most of them work in their line of specialization. It also shows that majority of them have been in the entity for more than a decade. However, it also turned out that there is a 9%

figure identifying some staff to be underqualified. It is worth noting that the strength of the team is on their post graduate qualification which is 60% of the total staff.

First: Descriptive study

This section analyzes the results of the study hypotheses under the main variables: 1) The responsibility of the auditor and his duties, 2) Audit report, 3) Independence, and 4) Audit evidence. This reflects the extent to which there is a gap in government audit expectations between government auditors, financial managers and beneficiaries of the government audit report.

The hypotheses and the analysis of the results will be discussed in three stages: in the first stage, the hypotheses was tested in the light of the main variables of the study, For the two-sample case, the researcher decided to use the permutation distribution, then the ranks of the data instead of the recorded observations was also considered. To determine acceptance or rejection of the hypothesis based on the results of the statistical analysis Kruskal-Wallis was used.

In the second phase, the non-linear, cross-sectional statistical analysis was tested for each statement separately to measure the direction of the responses of the parties involved in the study and to determine which statements give rise to the expectations gap for this variable. Finally, the Mann Whitney test was applied to bilateral comparisons between the responses of auditors, financial managers and beneficiaries to determine whether there are statistically significant differences between the responses of each group under each of the terms of the study.

The following are the mean and standard deviation, and average mean value for every paragraph at all variables:

Test the first hypothesis

Testing this hypothesis in paragraphs 1-6 of the questionnaire and Table 4 shows the overall results of the hypothesis test. The results of the extent to which there are gap of expectations among the three parties including in the study on the first major variable, "Responsibility of the Government Auditor and duties".

Table 4. The responsibility of government audit and duty.

Variable	average	Trends	k-w Chi(P-value)
Responsibility of government account audit and duty	69.88	Positive	**11.73=** x^2 **(0.003)** P-value

Table 4 shows the general trend of the responses of auditors, government employees and the legislature is significant about the terms that make up this variable. (69.88). This trend, however, reflects the results of Crocecal-Wallis (P-value 0.003, $x^2 = 11.75$) with significant differences between the responses of the three parties. Hence, the existence of the expectations gap in the government audit for the first major variable "the responsibility and duties of the government auditor". It is evident that using the Crosscal Wallace method, the test statistic a chi squared indicates 11.73, and the value of the P-value 0.003 is less than 0.05. So the hypothesis rejected and accepted null hypothesis. This means there are differences between respondents' responses to the study from the responsibilities and duties of the government auditor.

Table 5. Mean and standard deviation, and trends value for every paragraph

No.	Paragraph	Mean	Std. dev.	trends
1	Responsibility of government audit by control at all revenue and expenses	4.53	0.626	Accept
2	Responsibility of government audit by control at all assets	4.40	0.788	Accept
3	Responsibility of government audit is checking and auditing the accounts of public institutes and companies	4.55	0.624	Accept
4	Responsibility of government audit by control at monitoring the log in and out of the employee	4.31	0.633	Accept
5	Responsibility of government audit by investigating all penalties	4.34	0.649	Accept
6	Responsibility of government audit by fighting all types of fraud and corruption	4.30	0.788	Accept
Total mean for paragraph		4.40	0.684	Accept

It is noticed from Table 5 data that the ratio of the average of the group as a whole, is related to the responsibility of the government auditor and the duties. It was 4.40 and at the rate of 88%, and that the average result of the answers for all the paragraphs is close for all parties' knowledge of the auditor's responsibility and duties. The correct one indicates homogeneity of sample responses in all paragraphs.

Table 6. The responsibility of government account audit and duty

n	Paragraph	K-W	M-W: Multiple comparisons		
		Chi(P-value)	Sample (1) with sample (2)	Sample (1) with sample (3)	Sample (2) with sample (3)
1	The responsibility of government audit includes censorship on revenue and expenses	8.26 (0.016)	Z=-1.214 0.225	Z=-2.777 0.005	Z=-1.780 0.075
2	Responsibility of government audit by control at all assets	16.550 (0.000)	Z=-1.837 0.066	Z=-4.040 0.000	Z=-2.232 0.026
3	Responsibility of government audit by control at all accounts and institute	20.272 (0.000)	Z=-4.470 0.000	Z=-2.197 0.028	Z=- 2.104 0.035
4	Responsibility of government audit by control at absent and present	3.790 0.150	Z=-0.981 0.320	Z=-1.824 0.068	Z=-1.155 0.298
5	Responsibility of government audit by control at all plenty	19.836 (0.000)	Z=-4.195 0.000	Z=-1.458 0.145	Z=-3.139 0.002
6	Responsibility of government audit by control at all anti or avoid all corruption	2.480 (0.289)	Z=-0.347 0.729	Z=-1.494 0.135	Z=-1.205 0.228

To find out which of the six statements make up these variables which is attributed to the expectations gap in the government review, the Kruskal-Wallis test is done for each statement as shown in Table 6. The third term is "the responsibility of the government auditor to examine and audit public institutions and public companies." The most varied statements on which the opinions of the three parties participating in the study were 20,272 at a significant level of 0.000. This means that there is a very high indication in the gap, especially between members of the Bureau and the government sector. This is followed by the phrase "it is the responsibility of the government auditor to investigate violations" with a result of 19,836 at a significant level of 0.000. It is evident that there are no significant differences in paragraphs 4, 6, which provide for monitoring of the log in and log out of employees and fighting all types of fraud and corruption.

Test the second hypothesis

The second hypothesis was testing in paragraphs (7-9) of the questionnaire and Table (7) which shows the total results of the hypothesis test. The results measure the extent to which a gap of expectations exists between the three parties involved in the study on the second major variable, the Audit Report.

Table 7. The audit report.

variable	average	Trends	k-w Chi(P-value)
Audit report	13.11	Negative	**x2 = 0.760** **(0.684) P-value**

From Table 7 it is shown that using the Kruskal-Wallis test and the A chi squared test indicates 0.760, the value of the P-value of 0.684 is greater than 0.05, the null hypothesis was acceptable. There is no difference between respondent parties from the study about audit report.

Table 8. Mean and standard deviation, and trends value for every paragraph.

No.	Paragraph	Mean	Std. dev.	trends
7	I recognize that the report available	4.23	.719	Accept
8	The report dress all transparency	4.37	.636	Accept
9	Available for all people certain	4.22	0.704	Accept
Total mean for paragraph		4.40	**0.686**	Accept

Table 8 shows that the average of the group for the audit report is 4.27 and at the rate of 85%, and that the average result of the responses for each paragraph is close. The standard deviation of all the paragraphs is less than the sharp one, indicating the homogeneity of the sample responses around all the paragraphs. With the eighth highest in terms of acceptance with 4.37, indicating the majority's support that the government audit report is transparent.

Table 9. The audit report.

n	Paragraph	K-W	M-W: Multiple comparisons		
		Chi(P-value)	Sample (1) with sample (2)	Sample (1) with sample (3)	Sample (2) with sample (3)
7	I recognize that the report available	16.577 (0.000)	Z=-0.962 0.336	Z=-3.718 0.000	Z=-3.399 0.001
8	The report dress all transparency	12.443 (0.002)	Z=-1.006 0.315	Z=-3.536 0.000	Z=-2.289 0.022
9	Available for all people certain	6.429 (0.040)	Z=-1.820 0.069	Z=-2.257 0.024	Z=-0.866 0.386

Table 9, and through the use of the Kruskal-Wallis test for each of the constituent paragraphs of this axis, is noted that there are significant differences of paragraphs 7,8, and 9. This indicates that the total responses of the participants tend to agree to the variable. Accordingly, Mann Whitney's test of bilateral comparisons, there was no significant difference in the response of the participants.

Testing the third hypothesis

The third hypothesis was tested in paragraphs 10-12 of the questionnaire and Table 10 shows the total results of the hypothesis test. This measures the extent to which a gap of expectations exists between the three parties involved in the study on the third major variable which is the Independent.

Table 10. The Independent.

variable	average	Trends	k-w Chi(P-value)
Independent	17.090	Negative	**0.004 =x^2 p.value (0.998)**

From Table 10 it is shown by using the Kruskal-Wallis tested and A chi squared test of 0.004, the P-value of 0.998 is greater than 0.05, the null hypothesis will accept. There is no difference between respondent parties from the study about the independent variable.

Table 11. Mean and standard deviation, and trends value for every paragraph.

No.	Paragraph	Mean	Std. dev.	trends
10	Government audit has full finance independent	4.52	0.603	Accept
11	Government audit has full functional independent	4.38	0.681	Accept
12	Government audit has some pressure which prevent performance management	4.00	0.769	Accept
Total mean for paragraph		4.30	0.684	Accept

Table 11 shows that the average of the group for the independent is 4.30 and 86%, and that the average result of the responses for each paragraph is close. The standard deviation of all the paragraphs is less than the sharp one, indicating the homogeneity of the sample responses around all the paragraphs. With the tenth highest in terms of acceptance of 4.52, indicating majority support that the government's perfect independent

Table 12. The audit report.

n	Paragraph	K-W	M-W: Multiple comparisons		
		Chi(P-value)	Sample (1) with sample (2)	Sample (1) with sample (3)	Sample (2) with sample (3)
10	Government audit has full finance independent	0.707 (0.702)	Z=-0.713 0.476	Z=-127 0.899	Z=-734 0.463
11	Government audit has full functional independent	0.582 (0.748)	Z=-0.501 0.616	Z=-0.709 0.478	Z=-0.304 0.761
12	Government audit has some pressure which prevent performance management	9.266 (0.010)	Z=-2.092 0.036	Z=-1.147 0.253	Z=-3.024 0.002

Table 12 by Kruskal-Wallis test for each one separetly to know which one of the paragraph is expectations gap in government audit. There is number 12 which is much bigger significantly different in the response of the participants by 9.266 of P-value 0.010.

Test the Fourth hypothesis

The forth hypothesis was tested in paragraphs (13-17) of the questionnaire and Table (13) which shows the total results of the hypothesis test. The results measure the extent to which a gap of expectations exists between the three parties involved in the study on the fourth major variable, the audit evidence.

Table 13. The audit evidence.

Variable	average	Trends	k-w Chi(P-value)
Audit evidence	19.979	Positive	**x^2= 11.922 p.value (0.003)**

Table 13 as shown, that using the Kruskal-Wallis test and a chi squared test 11.922. Since the value of the P-value of 0.003 is greater than 0.05, the hypothesis is accepted. There is a difference between respondent parties from the study about audit evidence.

Table 14. Mean and standard deviation, and trends value for every paragraph.

No.	Paragraph	Mean	Std. dev.	trends
13	There is cooperate from other ministries with Audit Bureau Libyan	**3.97**	**0.759**	Accept
14	There is cooperate from other financial institute with Audit Bureau Libyan	**3.97**	**0.665**	Accept
15	Provide all data requested on time	**3.89**	**0.723**	Accept
16	Provide financial institute and public company full cooperation	**3.85**	**0.685**	Accept
17	There is evidence before provide the final report	**4.31**	**0.684**	Accept
Total mean for paragraph		3.31	0.703	Accept

Table 14 shows that the average of the group's for the audit evidence is 3.31 and 66%, and that the average result of the responses for each paragraph is close. The standard deviation of all the paragraphs is less than the sharp one, indicating the homogeneity of the sample responses around all the paragraphs. With the seventeenth highest in terms of acceptance of 4.31, indicating the majority's support that the government audit collected perfect audit evidence.

Table 15. The audit report.

n	Paragraph	K-W	M-W: Multiple comparisons		
		Chi(P-value)	Sample (1) with sample (2)	Sample (1) with sample (3)	Sample (2) with sample (3)

13	There is cooperate from other ministries with Audit Bureau Libyan	**2.787** **0.240**	Z=-0.717 0.473	Z=-1.059 0.290	Z=-1.678 0.093
14	There is cooperate from other financial institute with Audit Bureau Libyan	13.678 0.001	Z=-3.463 0.001	Z=-0.185 0.853	Z=-2.958 0.003
15	Provide all data requested on time	23.707 0.000	Z=-3.635 0.000	Z=-0.945 0.345	Z=-4.875 0.000
16	Provide financial institute and public company full cooperation	17.361 0.000	Z=-2.971 0.003	Z=-0.945 0.345	Z=-4.387 0.000
17	There is evidence before provide the final report	3.060 0.217	Z=-0.118 0.906	Z=-1.464 0.143	Z=-1.637 0.102

Table 15, through the use of the Kruskal-Wallis test for each of the constituent paragraphs of this table, is noted that there are significant differences of paragraphs 14,15, and 16. On the contrary to recognize which one was effective at gab audit government, Kruskal-Wallis test of each one was used. It turned out that no.15 is much bigger with significant difference in the response of the participants by 23.707 of P-value 0.000. This means of strong significance.

Conclusion

With the aforementioned analysis of the study the following is positied in this study of the development of budgeting in Libya:

1. There are gaps at government audit responsibility and duties as follows:

a. There is no gap between respondent parties to avoid all fraud and corruption;

b. There is big gap between government audit and government sectors about responsibility of auditing and inspection of institute and company;

2. There is no gap expectation for report audit;

a. The audit report dress all transparency for all government expenses and revenue;

b. There is no gap between three parties about the available for all citizen.

3. There is no forecast gap in the audit report, including the following:

a. It is clear that the report of the Audit Bureau expresses with transparency all the expenses of the state and its revenues;

b. There is no gap between the expectations of the three parties on the availability of citizens and its emphasis on the principle of transparency.

4. There is no gap in expectations regarding the independence of members of the Audit Bureau in the context of the following:

a. It is clear that the independence of the members of the Bureau, especially with regard to the pressures that hinder from performing the mission well, is the only one that may have a discrepancy with the government sector;

b. With regards to financial and functional independence, the arithmetic mean is higher than the pressures that hinder from performing the employees' duties.

5. There is an actual gap with respect to evidence of the following:

a. It is clear that there is a very high gap between the members of the Audit Bureau and the government sector regarding the cooperation of government departments and institutions with the SAI;

b. There is no gap between the members of the Audit Bureau and the government sector regarding the collection of sufficient evidence before writing the report.

Recommendation:

1. The public sector must be sensitized to the duties, responsibilities and duties of government authorities in accordance with local and international regulations and standards.

2. The necessity of activating the regulations concerning the provision of the government sector to all the accounting data of the Bureau in a timely manner and to provide full cooperation.

3. Training and continuous rehabilitation of the visitors to keep up with the latest developments and challenges facing them in the performance of their profession.

References

Chowdhury,R ,Jnnes , J and Kouhy , R. (2005)' The Public Sector Audit Expectation Gap in Bangladesh', *Managerial Auditing Journal*, 20(8) :893-908.

Chye, K. H and Woo, E.S. (1998)' The expectation gap in auditing' *Managerial Auditing Journal,* 13(3): 147-154.

Gaffney, M. and Lynn, S. (1989) 'The Expectations Gap and Municipal Auditing', *Government Accountants Journal*, 38(2):17-24.

Goldman, A and Benzion, B. (1974)' The Auditor- Firm Conflict of Interests: Its Implication for Independence', *The Accounting Review*, 49(4): 707-718.

Liggio, C.D. (1974) 'The expectation gap: the accountant's waterloo', Journal of contemporary business, 3, Spring:27-44.

John E. M and Stanley C. M. *(2001)* 'Auditors' and Investors' Perceptions of the "Expectation Gap", *Accounting Horizons*,15(4)

Olagunju, A. & Leyira, M. C., (2012) 'Audit expectation gap: perspectives of auditors and audited account users', International journal of development and management review,7(1).

Entrepreneurship policy for transforming local contexts in a developing economy: Evidence from Oman

Abdullah Al Shukaili *
University of Nizwa – Oman

Melih Madanoglu
Florida Atlantic University and Deusto Business School, Universidad de Deusto – Spain

José L. González-Pernía
Deusto Business School, Universidad de Deusto – Spain

Iñaki Peña Legazkue
Deusto Business School, Universidad de Deusto – Spain

Corresponding author
Abdullah Al Shukaili
Nizwa - Oman
(+968) 98991881, a.alshukaili@unizwa.edu.om

Abstract

It has been demonstrated in the literature that entrepreneurship contributes to local economic growth and development, and this evidence emanates mostly from advanced economies. Transforming local communities need path-breaking firms and pioneering entrepreneurs. A major goal of our research work is to gain insights on the effectiveness of entrepreneurship policy to support new firm growth in the milieu of a disadvantaged region. We build on previous work on entrepreneurship policy and institutional economics theories to hold that *direct* government support programs (i.e., small business loans and business training programs) may be necessary, yet insufficient, policy instruments to encourage entrepreneurship and new firm growth in local communities from developing economies. Our empirical work was conducted in Oman and data was collected for the period 2002-2011 from a database created by the Oman Development Bank (ODB). The database included information for the beneficiaries of the government financial program known as SANAD. Findings suggest that human capital constraints limited the growth potential of new ventures that benefited from the SANAD loan support program.

Keywords

Entrepreneurship policy, human capital, qualitative comparative analysis, developing economies

---

1 Introduction

Entrepreneurship is widely regarded as an engine for change and prosperity. The stimulus of entrepreneurial activity is particularly important in local communities of developing economies as new firms lead frequently to the emergence of new markets and modern social habits. In this paper, we claim that the impact of entrepreneurship could be more prominent in developing regions (i.e., regions where market and institutional failures abound and are more profound) if policies were more effectively implemented and monitored by local government authorities. It is widely accepted that to achieve economic and social goals, policies must be effectual in spurring *productive* entrepreneurship (Audretsch & Beckmann, 2007). Several studies have questioned the usefulness of public support programs for the promotion of *productive* entrepreneurship (Pena, 2004; Santarelli & Vivarelli, 2002). While the effectiveness of entrepreneurship policy for venture growth has been broadly investigated in the context of developed countries (Pergelova & Angulo-Ruiz 2014; van Praag & Versloot 2007), evidence remains scarce about the policy impact on new venture growth in less developed contexts (Park & Bae 2004).

Research issues like why publicly subsidized new firms fail to grow in developing economies have been widely ignored in the entrepreneurship policy literature. In an attempt to fill this gap, the purpose of our study is not to analyse whether or not entrepreneurship policy results in new firm growth in a developing economy. Rather, the goal is to disentangle the reasons behind the stagnation of new ventures, which benefit of entrepreneurship policies aimed at reinventing a local community in the context of a developing economy. Three are the main contributions of our study. Firstly, our findings shed light on the factors that prevent new firms from growth, despite the financial and business training support received from public institutions. Secondly, we delve into the human capital profiles of subsidized new-firm owners who fail to achieve venture growth by applying a barely used method in the entrepreneurship policy literature, namely the Qualitative Comparative Analysis (QCA) method. And lastly, our study provides evidence about the effectiveness of government support programs and delivers implications for policymakers to improve the design of entrepreneurship support programs and to transform local communities.

The remainder of this study is organized as follows. Section 2 provides an overview of the current debate about the role of government support programs to promote entrepreneurship in local contexts of developing economies. Section 3 describes the methodology and data applied in our empirical test. Results are briefly discussed in Section 4. The study ends with main conclusions and implications.

2. ENTREPRENEURSHIP POLICY TO ENCOURAGE NEW FIRM GROWTH IN DEVELOPING ECONOMIES

2.1 Government support programs to deal with the *"liability of newness"* of entrepreneurs

The high failure risk attributed to new firms stems from the newness of the business and the legitimacy needed to succeed in uncertain market environments. An important reason for government intervention to support new ventures is the so-called *"liability of newness"* faced by young companies (Stinchcombe, 1965). Most entrepreneurs face obstacles in accessing sources for funding and for meeting new customers, suppliers, and other stakeholders (Wiklund, Baker, & Shepherd 2010). Government subsidies are important support mechanisms to respond to weaknesses derived from the *"liability of newness"* (Söderblom, Samuelsson, Wiklund, & Sandberg, 2015).

The success level attained by local government institutions in encouraging *productive* entrepreneurship varies across economies. An orthodox view of this issue would suggest that *productive* entrepreneurship helps in transforming local communities, and therefore, institutional support is a well- accepted mechanism to promote *productive* entrepreneurship. Following this rationale, local communities could advance towards a wealthier stage through *productive* entrepreneurs supported by public institutions. This leads us to consider that local government agents should adopt different entrepreneurship policy mechanisms (i.e., such as innovation incentives, financing, taxation, trade and labour market regulations, etc.) which have worked successfully elsewhere and are suitable for each local context (Minniti, 2008). But as Mair & Marti (2009) hold, "one of the most prominent factors that prevent many developing countries from advancing along the road towards a market economy is the nature of their institutional fabric" (p.420). The limited evidence about the impact of the entrepreneurship policy on new business creation, survival and growth motivates us to further examine this issue in developing contexts, where both market failure and institution failure are more widespread and profound across local communities.

2.2 Policies to handle liquidity constraints faced by new firm owners

The existence of imperfect financial markets and information asymmetries between entrepreneurs and lenders have resulted often in government intervention to reduce and eliminate barriers that hamper entrepreneurship (Levitsky, 1997; Zecchini & Ventura, 2009). Banks normally provide loans to new firms at elevated interest rates, because the insolvency risk of young ventures is high. Flexible government intervention may contribute favourably to diminishing the negative impact of market failures and asymmetric information faced by entrepreneurs in financial markets (Felsenstein et al. 1998; Minniti 2008). For instance, government loan incentives are aimed at relaxing liquidity constraints and enhancing venture performance when entrepreneurs lack physical assets (which serve as collateral for loans), legitimacy, and trust from lenders (Blanchflower & Oswald 1998; Green 2003; Colombo, Croce, & Guerini 2010; Roman, Congregado, & Millán 2013).

While this mixed evidence refers to the effect of policies to overcome liquidity constraints faced by entrepreneurs in advanced economies (i.e., such as the US, UK or Japan), the literature on this issue is rare in the context of developing economies. Although the problem of accessing funds exacerbates in developing countries, a reduced number of entrepreneurs receive public support to create and develop their businesses. In our study, instead of acknowledging the existence of liquidity constraints and the shortage of financing for entrepreneurs, we examine the profile of owners of financially-subsidized new firms to better understand why most of them still face barriers to make their ventures grow in a local context of a developing economy.

2.3 Policies to improve business skills of entrepreneurs

Human capital theory suggests that managerial skills of new firm owners (i.e., abilities acquired through formal education, job experience, business training, etc.) positively affect new firm performance (Grilo & Thurik, 2005; Forbes, 2005). Although there is evidence in support of a positive relationship between formal education and new business growth (Bates, 1995), the majority of findings support the opposite result or a non-significant effect of entrepreneurs level of education on venture growth (Acs & Armington, 2004; Murdock, 2012; Davidsson & Honig, 2003). Other authors claim that the managerial experience of entrepreneurs plays a more important role than formal education to explain venture performance (Acs & Armington, 2004; Haber & Reichel, 2007).

The evidence on this issue in developing contexts is scarce. There is an ample consensus in that there is a need to increase the managerial skills of people from developing economies, especially women, who have a desire of launching a new firm. Our objective is to analyse the reasons that inhibit the beneficiaries of multiple public programs to engage in *productive* entrepreneurship (i.e., entrepreneurial activity that increases sales and creates new jobs beyond self-employment).

3 DATA AND METHODOLOGY

3.1 The local context and the data

Our empirical work was conducted in Oman and data was collected for the period 2002-2011 from a database created by the Oman Development Bank (ODB). The database included information for the beneficiaries of the government financial program known as SANAD. This program aims at helping entrepreneurs in both creating and growing their new businesses. After designing a questionnaire, interviews were conducted with entrepreneurs who were recipients of the loan provided by the SANAD support program. A total of 42 responses for the region of Batinah were used for our study.

We elected the region of Al Batinah to examine the effect of entrepreneurship-policy designed and implemented from the government of Oman. A high population growth, an excessive dependence of government on oil and gas revenues, a limited economic diversification, and an increasing dependence on foreign labor have resulted in continuous political and economic disequilibrium (Williams, Knight, & Rutter, 2015). According to IMF, Oman's economy will need to create 45,000 new job positions each year to reduce unemployment, especially among the youngest segments of the population (Oxford Business Group, 2013). It is worth to mention that Al Batinah is the second largest region in Oman, after Muscat, in terms of population size (i.e., with 717,192 residents) and the highest youth population between ages of 20 and 29.

A crisp-set Qualitative Comparative Analysis (csQCA) technique was applied to obtain results (Rihoux & Ragin, 2009). Our approach is not variable-based but case-based which allows for using small samples (10 to 100) and obtaining multiple pathways which lead to the same outcome (Lisboa, Skarmeas, & Saridakis, 2015). QCA fits our purpose because our objective is to unravel human capital related attributes that prevent new firms from growth, and thereby remain stagnant, while they benefit from the SANAD financial support program.

3 RESULTS

The QCA analysis exhibits six paths or configurations, which provide different profiles of owners of subsidized-ventures located in a local community of a developing country. Each configuration reflects the set of sufficient causal conditions, the unique coverage[1], the raw coverage, and the consistency of each configuration that constitutes each solution. Following Ragin, (2008) and other studies (e.g. Fiss, 2011; Greckhamer, 2015; Judge, Fainshmidt, & Brown III, 2014), we denote the core causal conditions with the symbol ⊕ and their absence with the symbol ❍. Likewise, the symbol ⊕ indicates the complementary causal conditions, whereas the symbol ○ indicates their absence.

As it can be seen in **Table 1**, the *consistency value* of our model is 0.96. A model is considered to be informative when the *consistency value* is above 0.74 (Woodside, 2013). The *raw coverage value* of our model is 0.89, meaning that the system of paths explains 89% of the financially subsidized-new firms facing stagnation (i.e., lack of employment growth for 3 consecutive years after inception).

4.1 Results for owners of financially subsidized-firms "not-participating" in *public business-training programs*

The three configurations of models 1, 2 and 3 represent the profile of owners of companies that do not experience growth in employment. The three models have in common that in each sub-group of entrepreneurs, financially subsidized-firm owners do not participate in *business training programs* offered by local government agents. The configuration of model 1 shows a profile of a male owner without previous managerial experience, whereas model 3 exhibits a male owner with an advanced formal education. The second configuration illustrates the profile of an entrepreneur (either male or female) without managerial experience, but with an advanced formal education.

The outcomes from half of our QCA tests highlight at least two results. First, male owners with formal education but no business experience, and vice versa, are likely to run businesses that fail to grow. These owners have in common that they receive financial support from the government, but do not participate in *public business-training programs*. One plausible

[1] Unique coverage : "indicating which share of the outcome is exclusively explained by a certain alternative path"; Raw coverage: "indicating which share of the outcome is explained by a certain alternative path";
Solution coverage: "indicating how much is covered by the solution term",
Consistency: "The coefficients of consistency and coverage provide important numeric expressions for how well the logical statement contained in the QCA solution term fits the underlying empirical evidence and how much it can explain" see (Schneider & Wagemann, 2010)

Table 1: Configurations for "stagnant" firms receiving *public financial support*

	Solution for Al Batinah					
	1	2	3	4	5	6
Male	⊕		⊕	○	◯	
Education		⊕	⊕	○	○	○
Experience	○	○		⊕		⊕
Support	◯	○	○		⊕	⊕
Consistency	1.00	1.00	1.00	0.91	0.91	0.88
Raw Coverage	0.13	0.10	0.07	0.37	0.34	0.28
Unique Coverage	0.10	0.07	0.03	0.21	0.17	0.10
Overall Solution Consistency		**0.96**				
Overall Solution Coverage		**0.89**				

Note: Core condition are represented by ⊕ (presence) and ○ (absence); complementary causal condition by ⊕ (presence) and ○ (absence).

explanation of this result is that such business owners do not seek firm growth.

Alternatively, they may not be capable of leading their ventures towards a growth path because they lack the specific skills to manage a new business. *Business training programs* arranged by institutions are usually tailored to solve this limitation that affects the human capital of entrepreneurs, (i.e., especially female entrepreneurs of developing economies). We run additional QCA tests to address this issue and to find out the profiles of new-firm owners who participated in *business-training programs*.

4.2 Results for owners of financially subsidized-firms "participating" in *public business-training programs*

Like in the previous tests, the three configurations of models 4, 5 and 6 represent the profile of owners of companies that do not experience growth in employment. The profiles of models 5 and 6 have in common that financially subsidized-firm owners participate in *business training programs* offered by local government agents. They are also similar in that the owners of both profiles lack an advanced formal education (i.e., this may be precisely a reason to enrol *in a public business training program*). We distinguish two different features among those owners who lack an advanced formal education but participate in public business training programs: either they are female entrepreneurs or they lack business experience (regardless their gender). Entrepreneurs with such profiles fail to launch growing businesses (i.e., even though they benefit from both government loan support and public business training services).

Similarly, model 4 suggests that female owners of financially subsidized-firms who lack an advanced formal education, but accrue managerial experience, are likely to run stagnant firms. That is, the outcomes from models 4 and 5 suggest that female owners with a low-level of education, despite their managerial experience or *business-training in public programs*, fall short in their aspiration to make their financially subsidized-businesses grow.

4 DISCUSSION AND CONCLUSION

Six profiles of entrepreneurs (i.e., individuals with specific configuration of variables) were identified as predictors of post-funding venture stagnation. The *consistency value* of the model was 0.96 and the *raw coverage value* was 0.89 which means that the model explained 89% of cases for firm's that faced stagnation (i.e., lack of employment growth for 3 consecutive years after inception).

According to our results, the six profiles of firms leading to stagnation were as follows. The first profile reflected women who lacked high-school education, but they had work experience. In the second configuration, we identified female who lacked high-school education but they received business training support from the government. The third profile consisted of individuals (i.e., both female and male) with work experience, with training support from the government, but no high-school education. In profile four, we found male without work experience and without training support from government. In configuration five, we found individuals with at least high-school education, without work experience and without training support from government. The sixth and last profile described firms run by men with at least high-school education and without training support from government. In sum, findings suggest that the absence of formal education by women entrepreneurs (women who were beneficiaries of the loan support program in Oman) led to stagnation, regardless their prior work experience or other business training support received from the government. Moreover, our findings also suggest that stagnation may come from the low human capital or absence of business training support of individuals in general. Therefore, human capital constraints limited the growth potential of new ventures that benefited from the SANAD loan support program.

With these findings in mind, our research contributes to the ongoing debate regarding the effectiveness of entrepreneurship policy by assessing its suitability in contexts where notorious market and institutional failures prevail. An important implication for policy makers is that providing small business loans is a necessary but an insufficient condition to achieve new firm growth. Policies to reduce financial market imperfections (i.e., such as the SANAD government loan support program in Oman) may be ineffective if the skills and human capital attributes of the beneficiary entrepreneurs are not sufficient to stimulate venture growth. New venture growth in developing economies needs more attention from institutions not only for reducing market failures (i.e., financial market failure) but also for improving the human capital of entrepreneurs who are the target of their support programs.

5 References

Acs, Z. J., & Armington, C. (2004). *New Firm Survival and Human Capital* (Papers on Entrepreneurship, Growth and Public Policy No. 2004–14). Max Planck Institute of Economics, Entrepreneurship, Growth and Public Policy Group. Retrieved from http://ideas.repec.org/p/esi/egpdis/2004-14.html

Audretsch, D., & Beckmann, I. A. M. (2007). From Small Business to Entrepreneurship Policy. In D. Audretsch, I. Grilo, & A. Thurik, *Handbook of Research on Entrepreneurship Policy*. Edward Elgar Publishing. Retrieved from http://www.elgaronline.com/view/9781845424091.00007.xml

Bates, T. (1995). Self-employment entry across industry groups. *Journal of Business Venturing*, *10*(2), 143–156.

Blanchflower, D., & Oswald, A. J. (1998). What makes an entrepreneur? Evidence on inheritance and capital constraints. *Journal of Labor Economies*, *16*(1), 26–60.

Colombo, M. G., Croce, A., & Guerini, M. (2010). The effect of public subsidies on firms' investment-cash flow sensitivity: ephemeral or persistent? Presented at the Paper presented at the Pacific Rim Innovation Conference 2010, Melbourne, Australia.

Davidsson, P., & Honig, B. (2003). The role of social and human capital among nascent entrepreneurs. *Journal of Business Venturing*, *18*(3), 301–331.

Felsenstein, D., Fleischer, A., & Sidi, A. (1998). Market failure and the estimation of subsidy size in a regional entrepreneurship programme. *Entrepreneurship & Regional Development*, *10*(2), 151–165. https://doi.org/10.1080/08985629800000009

Fiss, P. C. (2011). Building better causal theories: A fuzzy set approach to typologies in organization research. *Academy of Management Journal*, *54*(2), 393–420.

Forbes, D. P. (2005). Managerial determinants of decision speed in new ventures. *Strategic Management Journal*, *26*(4), 355–366. https://doi.org/10.1002/smj.451

Greckhamer, T. (2015). CEO compensation in relation to worker compensation across countries: The configurational impact of country-level institutions. *Strategic Management Journal*. Retrieved from http://onlinelibrary.wiley.com/doi/10.1002/smj.2370/pdf

Grilo, I., & Thurik, R. (2005). Latent and Actual Entrepreneurship in Europe and the US: Some Recent Developments. *The International Entrepreneurship and Management Journal*, *1*(4), 441–459. https://doi.org/10.1007/s11365-005-4772-9

Judge, W. Q., Fainshmidt, S., & Brown III, J. L. (2014). Which model of capitalism best delivers both wealth and equality [quest]. *Journal of International Business Studies*, *45*(4), 363–386.

Levitsky, J. (1997). SME guarantee schemes: a summary. *FINANCIER-BURR RIDGE-*, *4*, 5–11.

Lisboa, A., Skarmeas, D., & Saridakis, C. (2015). Entrepreneurial orientation pathways to performance: A fuzzy-set analysis. *Journal of Business Research*. Retrieved from http://www.sciencedirect.com/science/article/pii/S0148296315005226

Mair, J., & Marti, I. (2009). Entrepreneurship in and around institutional voids: A case study from Bangladesh. *Journal of Business Venturing*, *24*(5), 419–435. https://doi.org/10.1016/j.jbusvent.2008.04.006

Minniti, M. (2008). The role of government policy on entrepreneurial activity: productive, unproductive, or destructive? *Entrepreneurship Theory and Practice*, *32*(5), 779–790.

Murdock, K. A. (2012). Entrepreneurship policy: Trade-offs and impact in the EU. *Entrepreneurship & Regional Development*, *24*(9–10), 879–893.

Oxford Business Group. (2013). *The Report: Oman 2013*. Retrieved from https://books.google.es/books?id=89yb8WpTZ80C&lpg=PP1&pg=PA35#v=onepage&q&f=false

Park, S., & Bae, Z.-T. (2004). New venture strategies in a developing country: Identifying a typology and examining growth patterns through case studies. *Journal of Business Venturing*, *19*(1), 81–105.

Pena, I. (2004). Business incubation centers and new firm growth in the Basque country. *Small Business Economics*, *22*(3), 223–236.

Pergelova, A., & Angulo-Ruiz, F. (2014). The impact of government financial support on the performance of new firms: the role of competitive advantage as an intermediate outcome. *Entrepreneurship & Regional Development*, *26*(9–10), 663–705.

Ragin, C. C. (2008). *Redesigning social inquiry: Fuzzy sets and beyond* (Vol. 240). Wiley Online Library. Retrieved from http://onlinelibrary.wiley.com/doi/10.1002/j.1662-6370.2009.tb00140.x/abstract

Rihoux, B., & Ragin, C. C. (2009). *Configurational comparative methods: Qualitative comparative analysis (QCA) and related techniques*. Sage. Retrieved from https://books.google.es/books?hl=en&lr=&id=sAcIYzgO3nkC&oi=fnd&pg=PR1&dq=QCA&ots=uX_J9a00tC&sig=-k_D69fTnUzsmxjcc5QN61jE_j0

Roman, C., Congregado, E., & Millán, J. M. (2013). Start-up incentives: Entrepreneurship policy or active labour market programme? *Journal of Business Venturing*, *28*(1), 151–175. https://doi.org/10.1016/j.jbusvent.2012.01.004

Santarelli, E., & Vivarelli, M. (2002). Is subsidizing entry an optimal policy? *Industrial and Corporate Change*, *11*(1), 39–52.

Schneider, C. Q., & Wagemann, C. (2010). Standards of good practice in qualitative comparative analysis (QCA) and fuzzy-sets. *Comparative Sociology*, *9*(3), 397–418.

Söderblom, A., Samuelsson, M., Wiklund, J., & Sandberg, R. (2015). Inside the black box of outcome

additionality: Effects of early-stage government subsidies on resource accumulation and new venture performance. *Research Policy*, *44*(8), 1501–1512.
Sørensen, J. B., & Stuart, T. E. (2000). Aging, obsolescence, and organizational innovation. *Administrative Science Quarterly*, *45*(1), 81–112.
Stinchcombe, A. L. (1965). Social structure and organizations. In J. G. March (Ed.), *Handbook of Organizations*. Chicago: Rand McNally. Retrieved from http://www.emeraldinsight.com/doi/pdf/10.1016/S0742-3322(00)17019-6
van Praag, C. M., & Versloot, P. H. (2007). What is the value of entrepreneurship? A review of recent research. *Small Business Economics*, *29*(4), 351–382. https://doi.org/10.1007/s11187-007-9074-x
Wiklund, J., Baker, T., & Shepherd, D. (2010). The age-effect of financial indicators as buffers against the liability of newness. *Journal of Business Venturing*, *25*(4), 423–437.
Williams, W., Knight, H., & Rutter, R. (2015). A Study of the Convergence between Entrepreneurship, Government Policy and Higher Education in the Sultanate of Oman. Presented at the 60th Annual ICSB World Conference, Entrepreneurship at a Global Crossroads, Dubai, UAE, June 6-9.
Zecchini, S., & Ventura, M. (2009). The impact of public guarantees on credit to SMEs. *Small Business Economics*, *32*(2), 191–206.

Tourism Entrepreneurship: a Path towards Discovery and Culture

Martina Juvara,
URBAN Silence, United Kingdom,
martinaj@urbansilenceltd.com

Oman's National Tourism Strategy has great ambitions to increase the role of tourism in the country, significantly increasing the number of visitors up to 11 million a year and offering high value and low volume 'experiences', rather than the resort holidays more common to other Gulf countries.

This Presentation will explore how entrepreneurship in tourism has created different models that lead to the discovery of the places visited and the valorisation of culture locally, while creating stable and long-term employment and increase in local skills. The objective will be to identify viable and replicable models of local entrepreneurship at local level that lead to respectful and culturally rich tourism experiences in a range of countries in the world, and consider which models could be relevant for Oman.

Indicatively, the examples could cover:

- The regeneration of the Walled City of Nicosia (Cyprus), in ruins until a couple of decades ago and now a thriving centre of youth culture, fashion and art and the most visited destination in the city, both by locals and visitors to the island.
- The revival (and reinvention) of traditional food and crafts into a high value offer in Siem Reap (Cambodia) through charitable organisations initially aimed at providing skills and jobs to destitute families and now making a major contribution to the 'outward presentation' of the country.
- The guided tours and experiences of the Sapa Valley trekking organisations (Vietnam), involving local people in presenting their own territory to visitors and making them appreciate the hardships and beauty of local life.

These examples, and others similar, started with a combination of local enterprise, public sector recognition and support (in most but not all cases), internet promotion. All maximise the return for the local economy (visitor spend, employment, skills) and increase the appreciation of local assets: geographical or cultural.

Each model will be categorised in term of typology of tourism and of enterprise, key actors/ managers and business model, spin-offs in term of employment, visitor numbers and cultural appreciation or development. As part of the exploration of the business model, there will be consideration of the required infrastructure and enabling investments, and where appropriate or relevant, the planning and legislative framework required. A brief comparative analysis will be made, and reference to the Omani National Tourism Strategy objectives will be used to identify which of these models could be an inspiration for Oman.

References

Keywords: *Tourism, Local Enterprise, Culture, National Tourism Strategy*

Stock splits: A review, reconceptualization, and extension

Bilal Ahmad Pandow
Research Scholar
Department of Commerce, University of Kashmir
Srinagar, India
ibilalhussain@gmail.com

Prof. Khurshid Ahmad Butt
Dean
School of Business Studies, University of Kashmir
Srinagar, India
kahmad2012@yahoo.com

Abstract**—***The stock split is a corporate action where companies divide its existing equity-shares into multiple stocks. A split is not expected to substantially affect a splitting company's financials. Also, the splitting corporates' existing shareholders continue to hold the same percentage holding in the company before and after a stock split. Despite these academic facets, a company's stock split declaration has been witnessed as an encouraging influential event. It has also been observed that after a split new investors get attracted in buying the stock as it is available at a lower price. Though, the stock split may have no impact on the value of the investment as the fundamentals of the company remain unchanged. The question arises do such investors actually end up gaining or losing? However, it is expected that the market forces determine the real price of the scrips as the liquidity would increase and more shares become available after the split. The questions like this highlights that it is significant for investors to comprehend the purpose and the aim of the firm going for the split. The present study will address these questions in the Indian context. The aim of the study is to examine two important aspects related to the stock split event: the effect on the liquidity of firms that go for a split and the impact on stock price. The study will focus and review the literature available in Indian and global context. Also, this study stems from the fact that not many studies are available in this specific corporate action context. Besides, this study intends to contribute to the existing literature by studying volatility and liquidity effects of stock split. In order to achieve the set objective of the study, the researcher shall make the use of existing studies from credible sources.*

Keywords**— stock split, corporate action, volatility, liquidity, return, shareholder**

I. Introduction

Researchers have used the stock splits to explain various corporate actions. In this article we review the literature to identify key reasons for stock splits and offer a reconceptualization of this corporate action. Building upon the dynamic capabilities view of the firm, we distinguish between a firm's potential and realized capacity.

A plethora of research work has been carried out on stock splits relating to liquidity hypothesis, trading range hypothesis, signaling hypothesis and neglected firm hypothesis. This paper offers a gist of the existing research undertaken on stock splits.

The study includes the review of various studies conducted on the related topic of the study by the researchers. This review will help in understanding many perspectives on the issue and enables the researchers to develop a suitable direction for the future research.

The researchers have developed a model of split behavior where the split serves as a costly signal of managers' reserved evidence due to costs of trading in stock which depends on stock prices [1]. Similarly, there have been investigation on how dynamic information acquirement improves price-based return movement, a typical variance in stock splits [2]. While the principal motives for the splits can be largely be characterized into two groups: those who contend that firms use the splits to indicate promising information to the market and those who argue that the splits are used to advance liquidity [3]. Also, other observed that the splits assumes that managers use stock splits to the market about the reasonable price of the firm and thus endeavor to diminish undervaluation [4]. Echoing similar views the researcher have found that the firms do expect higher volatility following splits [5].

II. Review and Reconceptulization of Stock Splits

The companies split stocks because of the following main reasons:

- Liquidity: By encourage more retail investors to trade.
- Signaling: By signaling about the future prospects of the company.
- Trading Range: By bringing the prices of the shares to the desired trading level so as to attract small traders.
- Neglected Firm issues: Drawing the attention of investors towards the company.

There are several theories developed in the literature to explain stock split behavior. The more popular theories are signaling theory, increased liquidity theory, optimal trading range theory and dispersion of control theory.

A. Liquidity

The researchers have examined the reaction of trading volume activity and security return variability to annual earnings announcement with a sample of 143 New York Stock Exchange firms. The results obtained indicated that a 33 percentage increase in trading volume and 61 percentage increase in security returns in earnings announcement week over the non-announcement week [6]. According to [7], in their study found that the announcement of a split sets off a chain of events such as an increase in the daily number of transactions which in turn increases the noise-ness of the security return process.

While, [8], analyzed the role of information in announcing of stock splits by Companies plays an important role in price determination. They have defined information as "Pure Information" (i.e., which contains details only about the stock splits) for the purpose of the study. The results have shown that the market anticipation of split announcements is related to the stock's information richness.

Also, there are studied [9] 1131 stock splits between1983-1989. There is an increase in the number of trades as well as a decrease in the mean trade size following the split. The number of liquidity trades also increases after a split. Further, changes in the numbers of individual and institutional shareholders are positively related to the split factor. Abnormal announcement returns are positively correlated with changes in the total number of shareholders. These findings support the signaling hypothesis [10].

There are studies [11] that have examined changes in trading activity around stock splits, and their effect on the volatility and the adverse information component of the bid-ask spread. Even after controlling for microstructure biases, it is found that there is a significant increase in the volatility after the split. Changes in total volatility and in its permanent component are positively related to changes in the number of trades. The study [12] has provided enough evidence that in the US, stock splits are associated with positive abnormal returns around the announcement and the execution day. They found excess returns of 3.4 percent on the announcement date.

Companies tend to split their stock so that the institutionally mandated minimum tick size stays in the optimal percentage range of stock price. A larger tick size increases incentives for dealers to make markets and for investors to provide liquidity through placing limit orders, although at a cost of an increased minimum bid-ask spread. A simple model demonstrates the correlation between firm size and share price. A minimum price variation of $.01 in the U.S. could eventually lead through stock splits to an average share price around $3.00 [13].

The study [14] examine the "marketability hypothesis," which states that stock splits enhance the attractiveness of shares to investors by restoring prices to a preferred trading range. They found that splitting funds experience significant increases (relative to non-splitting matched funds) in net assets and shareholders. Stock splits do appear to enhance marketability.

The study [15] used non-public NYSE system data to study the effect of stock splits on liquidity by documenting changes in the limit order book, execution costs, and trading activity. They find that depth available in the limit order book at various dollar distances from the mid-quote increase after a stock split, but the depth available at a various percentage (split-adjusted) distances declines substantially.

The study [16] has tried to prove the effect of stock splits on the brokers in NYSE, the period of study being between 1993 and 1994. The study finds that the brokers promote all stock split stocks which give wider bid-ask spread in turn more margins. The study [17] has reexamined long-run performance following stock splits during 1950 to 2000. Significantly positive and robust equally weighted abnormal returns are documented during the first year following the announcement month; however, significant value weighted long-run abnormal returns are largely confined to the period from 1975 to 1987. When long-run performance is examined following the ex or effective date of stock splits, abnormal returns are insignificant, except for equally weighted portfolios during 1975 to 1987. Further analysis documents that the equally weighted long-run abnormal performance during 1975 to 1987 is strongly correlated with unexpected decreases in post-split systematic risk.

The study [18] tried to test small trader hypothesis. According to small trader hypothesis, small traders prefer low priced stocks and the post-split volatility increase may be due to the trading activity of these small traders. The research work has proved that there are significant effects of trading activity by small traders on the post-split increase in volatility.

Also, there are studies [19] examined the relationship between institutional investors and stock split behavior of firms between 1994 and 1999. The author has also made an attempt to test certain hypotheses about institutional ownership effects upon firm split behavior, signaling hypothesis, and the association between split behavior and pre-existing institutional ownership levels etc. According to [20] the relation between trading activity, return volatility and auto correlations around stock splits were studied. Also, the researchers studied [21] the effect of stock splits on the liquidity of stocks. They chose around 2500 stock splits and found a pervasive decline in most measures. Large stock splits exhibit a more severe liquidity decline than small stock splits, especially on NASDAQ.

A possible explanation of these related phenomena is that trading hindrances associated with record dates create trading inconvenience that is reflected in lower prices near record dates. In turn, anomalous positive ex-date returns arise in part from the abnormally low prices of un-split shares caused by the negative record date returns [19].

Besides, there are studies [22] that examined shareholder wealth and volatility effects around the announcement and execution dates of stock splits at the Helsinki and Stockholm Exchanges. The empirical study of the paper provides abnormal returns around the announcement and execution dates of stock split at the Stockholm Exchange but not at the Helsinki Stock Exchange. The study [23] have presented a microstructure of stock splits in the presence of minimum tick size rules. The key feature of the model is that discretionary trading is endogenously determined. The research [24] discuss that academic and practicing accountants use forecasts made by security analysts to estimate market expectations about forthcoming earnings announcements.

The researchers in their study [25], the authors have identified that stock return volatility tends to increase significantly following stock splits. The cause for this is the trading of stocks in discrete price intervals called ticks. The study was to analyze whether there is stock return volatility of split stocks after a reduction in the tick size from 1/8th to 1/16. The results show that post return variance increase was unaffected by the reduction in tick size, thus concluding that price discreteness is not a determinant of affecting the change in returns following stock splits.

According to [26] the interest in short selling decreases when stock splits show a favorable signal about the future performance of firms and short selling increases when firms expect more post-split liquidity improvements.

The study [27] have analyzed the effect of stock splits on liquidity. The study has also found a change in the trading composition with an increase in the smallest transactions, mainly of the shares whose prices fall significantly after the split. Finally, despite higher transaction costs, splits encourage small investors because of lower stock prices.

Also, there were efforts made to distinguish signaling hypothesis from liquidity hypothesis [28]. The literature survey shows the author has mixed results as far as signaling and liquidity hypothesis are concerned. The firms which opt for stock splits generally send a signal that the future earnings are going to be incredulous and hence this assumption will encourage small investors to buy more of such stocks, thus enabling more liquidity. The study [29] made an attempt to study the performance of 12747 stock splits from 1927 to 1996 using size and book-to-market variables to measure abnormal returns. The study shows that only small or negligible abnormal returns the sample earn using the calendar-time method.

The researchers examined the market behavior surrounding stock split announcements in the Canadian market for the 1977-1993 period and the effect of 2 years before compared to 2 years after the announcement. Using the event study methodology, the findings indicate that positive abnormal returns (AR) exist on both the announcement days (0, 1) and the 11-day period surrounding stock split announcements. The results also show that following the split event, bid-ask spreads decrease, while both trading volume and the number of transactions increase thus suggesting that split events enhance liquidity. Further, it appears that earnings grow in the 2 year period following split events, thus implying that split events signal future performance of the firm [30].

B. *Signaling hypothesis*

The occurrence of stock split doesn't affect cash flows of the firm and there is no change in the capital structure of the company, so the net worth of the firm is not be affected by the stock split. However, the empirical research on splits by [31], [32], and [33] find abnormal return around announcement day which is explained as the support to the hypothesis of split announcement effect. Not making use of the event of a stock split and adapting the signaling model of [34], [34], [35] and [37] propose that financial decisions of the management convey information about the firm value.

One need not get confused over signaling hypothesis with that of attention getting hypothesis; there is a clear-cut distinction between the signaling hypothesis and attention-getting hypothesis which states that the reaction to stock will be positive results if the company will draw more attention from the market.

The valuation effects of stock splits and stock dividends are studied by [13]. The study concluded that the information of stock split and dividend appears to be associated with firms' future cash flows, future earnings or future worth of equity.

A study conducted by [7] observed the abnormal return and trading volume after the stock split. The study found a positive abnormal return and increase in the volume of shares traded after the announcement which supports the signaling hypothesis that forecast the positive market reaction to the stock split.

Also, in support of the signaling hypothesis, [38] conduced a test on whether stock dividends and stock splits conveyed the information about company's future earnings and also test whether the split factor acts as a signal.

The assumption that trading is costly, based on this [16] developed a model of stock-split behavior in which the split serves as a costly signal of managers' private information because stock trading costs depend on stock prices. The findings of [39], [40], and [41] also find supports the signaling hypothesis that managers of undervalued firms communicate information to their investors by splitting their stocks.

The positive signal estimation is not always credible taking the stock split's abnormal return as evidence, as the abnormal return can also be due to the improved liquidity after the stock price decreases. The study [42] found an alternative way to make the signal test. Also, the signaling hypothesis cannot explain why excess stock returns are also observed around the split date as the ex-split date is predictable following the split announcement, and there shall be no information content in the split at the ex-date. So researchers give other explanations for the stock split. Similarly, [13], [43] explain excess returns from the viewpoint of microstructure and suggest the liquidity-based explanation for stock splits, and this introduces the second hypothesis.

C. *Trading Range*

The study [44] claims that companies prefer keeping the price of their stocks within a certain range. The researcher explains it by the argument that the management tries to keep stock prices within a certain range and the reason is to attract a certain kind of clientele or to scatter the ownership of the company. The split is considered to attract the small investors who get enticed by the lower price. The researchers conducted a survey [45] on management's view of the stock split. The survey reveals that majority of financial analysts of various companies believe that the stock split is a worthy tool to bring the stock into an optimal trading range.

Evidence have been found to support this hypothesis, [7] found the daily number of transactions along with the raw volume of shares traded increase after the split. The research [46] made use of the trading volume to measure the small and large traders' activities and fond that small traders transact more frequently after the stock split and the trade direction changes significantly from sell to buy. The study [47], [48], and [16] also supported the trading behavior change after the stock split.

However, why companies want to attract the small or uninformed traders to invest in their stocks? One of the most popular explanations is that the enlarged investor base will increase the liquidity of the stock. Liquidity refers to the frequency and ease with which an asset can be converted to cash. The study [49] defines it as "accommodation of trading with the least effect on price". The motivation of managers to split stocks is to enhance the liquidity. From the viewpoint of liquidity, the trading range hypothesis is also called "liquidity hypothesis".

The liquidity enhancement is explained by the reasoning that the lower price after stock split will attract small investors who are always thought as wealth-constrained. The participation of more investors increases the liquidity of the stock. The study [50] sets up a model of capital market equilibrium with incomplete information. The study [51] states that the noise trading is essential to the existence of liquid markets, and the more noise trades there are, the more liquid the markets will be.

For the computation of the liquidity, we need proxies and the proxies are classified into two categories: measures of friction and activity that are the two dimensions of liquidity. The friction measure is defined by [52], [53], and [54] as the price concession needed for an immediate transaction. Friction in financial markets measures the difficulty with which an asset is traded.

The study [55] takes the monthly shares traded relative to the shares outstanding as the measure to stock's marketability and finds that the result does not support a decrease in marketability though stock split's effects on the trading volume are not permanent.

Also, the research [56] observed the trading patterns of small and large traders around stock split ex-dates. Using the intraday transaction database for the Toronto Stock Exchange during 1983-89, researchers found that stock splits are associated with significant changes in trading patterns. The findings suggest that stock splits are mainly aimed at restoring stock prices to a 'normal range'. The study [57] suggested that stock splits should have no effect on firm value in perfect capital markets, yet stock prices increase on split announcements.

The researchers found that there is an increase in small traders' participation after stock splits that result in the increase of liquidity by adjusting values of the liquidity premium defined as the absolute value of the difference between the transaction price and the average of the bid and ask price following the method [58].

The study [59] also examined the effects of stock splits during the period between 1st January 1981, and 30th April 1983, on bid-ask spreads for NYSE-listed companies. They found that percentage spreads increase after splits, representing a liquidity cost to investors. These spread increases are directly related to decreases in share prices following splits and can explain part, but not all, of the observed increase in return variability after splits. The evidence thus suggests a liquidity cost of stock splits that must be weighed against any other perceived benefits of splits. Such a liquidity cost may validate that stock splits are a signal of favorable information about the firm.

Besides, the paper [60] analyzes the empirical behavior of stock-return volatilities prior to and subsequent to the ex-dates of stock splits. Conroy et al. (1990) studied effects of stock splits on bid-ask spreads for NYSE-listed companies. They find both the percentage spread and daily price-return variance increases and the regression result shows that part of the split-induced increase in price-return variance can be attributed to bid-ask spreads. And their result is consistent with [61] concluded that 'a partial (but not complete) explanation of the volatility increase' is to be found in institutional factors such as the bid-ask spread. They couldn't conclude that the increase of the price-return volatility is due to smaller traders' participation although empirical results do not support the hypothesis that the inflow of smaller traders is good for market stability.

While the study conducted by [62] found the volatility change subsequent to NYSE and AMEX stock splits. Three main differences exist between the two stock exchanges: attributes of the stock and/or company, special behavior, and ownership clientele factors. The study [63] use the event of when-issued shares which are traded at the post-split price level before the split to study noise traders or small traders' trading behavior. They find the volatility of shares before the split is lower for both the un-split shares and the when-issued shares by comparing with matching firms. After the split, the volatility measure increases significantly and they explain this phenomenon is due to the small-volume traders' return to trading in a regular way with a single price level.

However, the fact is that both uninformed and informed traders will take part in trades more aggressively after stock splits and that is what [64] argue. Their model shows that in the context of stock splits if lower post-split shares' price attracts noise traders, the level of informed traders' participation will increase endogenously.

The study [11] argue that changes in trading activity will affect adverse-information component of the bid-ask spread which is decomposed into three components: order processing, adverse information, and inventory cost [65]. They adopt the method proposed by [66] to decompose the total spread into the order-processing and adverse-information components. They found the adverse-information component increases by 0.17 or a 22% increase over the pre-split value.

D. Neglected Firm

The regression analysis suggests that the positive reaction can be attributed to the Neglected firm hypothesis and liquidity hypothesis [67]. There is evidence against signaling hypothesis as well as neglected firm hypothesis. There is evidence of positive abnormal return on ex-split date but it reverses soon after. And therefore, there is also evidence against liquidity or trading range hypothesis [68].

The sample stocks has shown increasing mean returns after the split, and the "neglected firm" hypothesis apply in this case with the stock being relatively thinly traded before the split [69].

There is a positive and significant impact of stock splits on the financial ratios like earnings per share, return on equity, profit per share and price to earnings ratio [70].

The results found in Indian contest during this study doesn't provide any conclusive evidence about positive abnormal return associated with announcement of stock split which also discards signaling hypothesis and neglected firm hypothesis presented [71].

Split-induced returns cannot be explained by variables proxy for conventional liquidity and signaling hypotheses for stock-split activity. Positive split-induced returns are largely reversed in the post-split months. Post-split stock underperformance is inversely related to, and thus appears to be a correction for, the significant market overreaction at split execution. They suggest an investor irrationality explanation for these results, arguing that stock splits were associated with the creation of the bubble due to the inability of investors to understand splits correctly. We conclude that educating investors in emerging markets to process information correctly will improve the efficiency of such markets [72].

III. Extension

In past, many empirical studies have been conducted to establish the impact of stock splits on price and liquidity across different countries. There have been many studies conducted in U.S. and other developed countries context. Although market reacts differently to stock splits in U.S., the methodology of such studies has helped to develop an effective theory for the study in the emerging economic particularly in Indian context. As the subject has not received much attention amongst researchers in many of the emerging economies especially countries like India, there is very little understanding on the effects of stock splits in the Indian context. Therefore, the study in emerging economies particularly in Indian context should be conducted to analyze and interpret the effect of a stock splits.

References

[1] Brennan, M.J. and Copeland, T.E., 1988. Stock splits, stock prices, and transaction costs. *Journal of Financial Economics*, 22(1), pp.83-101.

[2] Kong, D., Lin, C. and Liu, S., 2017. Does Information Acquisition Alleviate Market Anomalies? Categorization Bias in Stock Splits. *Review of Finance*.

[3] Li, F., Liu, M.H. and Shi, Y.E., 2017. Institutional ownership around stock splits. *Pacific-Basin Finance Journal*, 46, pp.14-40.

[4] Karim, M.A. and Sarkar, S., 2016. Do stock splits signal undervaluation?. *Journal of Behavioral and Experimental Finance*, 9, pp.119-124.

[5] Gharghori, P., Maberly, E.D. and Nguyen, A., 2017. Informed trading around stock split announcements: Evidence from the option market. *Journal of Financial and Quantitative Analysis*, 52(2), pp.705-735.

[6] W. H. Beaver, "The Information Content of Annual Announcements Earnings," *J. Account. Res.*, vol. 6, no. 1968, pp. 67–92, 1968.

[7] Lamoureux and Poon, "The Market Reaction to Stock Splits," *J. Finance*, vol. 42, no. 5, pp. 1347–1370, 1987.

[8] A. Arbel and G. Swanson, "The Role of Information in Stock Split Announcement Effects in Stock Split Announcement The Role of Information Effects," *Q. J. Bus. Econ.*, vol. 32, no. 2, pp. 14–25, 1993.

[9] S. P. Ferris, C. Y. Hwang, and A. Sarin, "A microstructure examination of trading activity following stock splits," *Rev. Quant. Financ. Account.*, vol. 5, no. 1, pp. 27–41, 1995.

[10] S. Mukherji, Y. H. Kim, and M. C. Walker, "The effect of stock splits on the ownership structure of firms," *J. Corp. Financ.*, vol. 3, no. 2, pp. 167–188, 1997.

[11] A. S. Desai, M. Nimalendran, and S. Venkataraman, "Changes in Trading Activity following Stock Splits and their Effect on Volatility and the Adverse-Information component of the Bid-Ask Spread," *J. Financ. Res.*, vol. 21, no. 2, pp. 159–183, 1998.

[12] J. L. Koski, "Measurement effects and the variance of returns after stock splits and stock dividends," *Rev. Financ. Stud.*, vol. 11, no. 1, pp. 143–162, 1998.

[13] M. S. Rozeff, "Stock splits: Evidence from mutual funds," *J. Finance*, vol. 53, no. 1, pp. 335–349, 1998.

[14] C. Fernando, S. Krishnamurthy, and P. Spindt, "Is share price related to marketability? Evidence from mutual fund share splits," *Financ. Manag.*, vol. 28, no. 3, pp. 54–67, 1999.

[15] C. M. Jones, G. Kaul, and M. L. Lipson, "Transactions , Volume , and Volatility Reproduced with permission of the copyright owner . Further reproduction prohibited without permission .," *Rev. Financ. Stud.*, 1998.

[16] P. Schultz, "Stock Splits, Tick Size, and Sponsorship," *J. Finance*, vol. 55, no. 1, pp. 429–450, 2000.

[17] R. D. Boehme, "Reexamining The Long-Run Stock Split Anomaly Puzzle," *SSRN Electron. J.*, no. November, 2001.

[18] C. Chen, "Small traders and increased price volatility after stock splits," Syracuse University, 2001.

[19] M. S. . Nayar, Nandkumar; Rozeff, "Record Date , When-Issued , and Ex-Date Effects in Stock Splits," *J. Financ. Res.*, vol. 36, no. 1, pp. 119–139, 2001.

[20] A. Kamara and J. L. Koski, "Volatility, autocorrelations, and trading activity after stock splits," *J. Financ. Mark.*, vol. 4, no. 2, pp. 163–184, 2001.

[21] D. Michayluk and P. Kofman, "Liquidity, Market Structure, and Stock Splits," *Work. Pap.*, vol. July, pp. 1–37, 2001.
[22] N. Antti, "Shareholder Wealth and Volatility Effects of Stock Splits Some Results on Data for the Helsinki and Stockholm Stock Exchanges," *Liiketal. Aikakausk.*, vol. 49, no. 1, pp. 37–70, 2001.
[23] V. R. Anshuman and A. Kalay, "Can splits create market liquidity? Theory and evidence," *J. Financ. Mark.*, vol. 5, no. 1, pp. 83–125, 2002.
[24] W. R. Baber and S.-H. Kang, "Is Meeting the Consensus EPS Good News or Bad News?," *no.* 202, 2002.
[25] W. Y. Lee, C. X. Jiang, and D. C. Indro, "Stock market volatlity, excess returns, and the role of investor sentinent," *J. Bank. Financ.*, vol. 26, no. 12, pp. 2277–2299, 2002.
[26] Kadiyala and Vetsuypens, "Are Stock Splits Credible Signals? Evidence from Short-Interest Data," *Financ. Manag.*, vol. 31, no. 1, pp. 31–49, 2002.
[27] Guirao and Sala, "Transaction Size, Order Submission and Price Preferences Around Stock Splits," *SSRN Electron. J.*, no. June, 2002.
[28] P. Dennis and D. Strickland, "The effect of stock splits on liquidity and excess returns: Evidence from shareholder ownership composition," *J. Financ. Res.*, vol. 26, no. 3, pp. 355–370, 2003.
[29] Byunl and Rozeef, "Long□run Performance after Stock Splits: 1927 to 1996," *J. Finance, vol.* 58, no. 3, pp. 1063–1085, 2003.
[30] S. Elfakhani and T. Lung, "The effect of split announcements on Canadian stocks," *Glob. Financ. J.*, vol. 14, no. 2, pp. 197–216, 2003.
[31] E. F. Fama, L. Fisher, M. C. Jensen, and R. Roll, "the Adjustment of Stock Prices To New Information.," *Int. Econ. Rev. (Philadelphia).*, vol. 10, no. 1, p. 1, 1969.
[32] S. Bar□ Yosef and L. D. Brown, "a Reexamination of Stock Splits Using Moving Betas," *J. Finance*, vol. 32, no. 4, pp. 1069–1080, 1977.
[33] G. Charest, "Dividend information, stock returns and market efficiency-II," *J. financ. econ.*, vol. 6, no. 2–3, pp. 297–330, 1978.
[34] M. Spence, "Job Market Signaling," *Q. J. Econ.*, vol. 87, no. 3, pp. 355–374, 1973.
[35] S. S. A. Ross, "The determination of financial structure: the incentive-signalling approach," *Bell J. Econ.*, vol. 8, no. 1, pp. 23–40, 1977.
[36] H. E. Leland and D. H. Pyle, "Informational Asymmetries, Financial Structure, and Financial Intermediation," *J. Finance*, vol. 32, no. 2, pp. 371–387, 1977.
[37] S. Bhattacharya, "Imperfect Information, Dividend Policy, and 'The Bird in the Hand' Fallacy," *Bell J. Econ.*, vol. 10, no. 1, pp. 259–270, 1979.
[38] M. M. and A. Dravid, "Stock Dividends , Stock Splits , and Signaling," *J. Finance*, vol. 45, no. 3, 1990.
[39] Brennan and Hughes, "Stock Prices and the Economy," *J. Bus.*, vol. XLVI, no. 5, 1991.
[40] E. K. S. David L. Ikenberry, Graeme Rankine, "What Do Stock Splits Really Signal□?," *J. Financ. Quant. Anal.*, vol. 31, no. 3, pp. 357–375, 1996.
[41] R. M. Conroy, R. S. Harris, and S. Robert, "Stock Splits and Information: The Role of Share Price," *Financ. Manag.*, vol. 28, no. 3, pp. 28–40, 1999.
[42] Kadiyala and Vetsuypens, "Are Stock Splits Credible Signals? Evidence from Short-Interest Data," *Financ. Manag.*, vol. 31, no. 1, pp. 31–49, 2002.
[43] Maloney and Mulherin, "The Effects of Splitting on the Ex: A Microstructure Reconciliation," *Financ. Manag.*, vol. 21, no. 4, pp. 44–59, 1992.
[44] T. E. Copeland, "Liquidity Changes Following Stock Splits," *J. Finance*, vol. 40, no. 1, pp. 309–317, 1979.
[45] Baker and Gallagher, "Management's View of Stock Splits," *Financ. Manag.*, vol. 9, no. 2, pp. 73–77, 1980.
[46] Kryzanowski and Zhang, "Trading Patterns of Small and Large Traders Around Stock Split Ex-Dates," *J. Financ. Res.*, vol. XIX, no. 1, pp. 75–90, 1996.
[47] Angel, Brooks, and Mathew, "when-issued shares, small trades, and the variance of returns around stock splits," *J. Financ. Res.*, 2004.
[48] A. S. Desai, M. Nimalendran, and S. Venkataraman, "Changes in Trading Activity following Stock Splits and their Effect on Volatility and the Adverse-Information component of the Bid-Ask Spread," *J. Financ. Res.*, vol. 21, no. 2, pp. 159–183, 1998.
[49] M. O'Hara, "Overview□: market structure issues in market liquidity," *BIS Pap.*, no. 2, pp. 1–8, 1995.
[50] R. C. Merton, "A Simple Model of Capital Market Equilibrium with Incomplete Information," *J. Finance*, vol. 42, no. 3, pp. 483–510, 1987.
[51] F. Black, "Noise," *J. Finance*, vol. 41, no. 3, pp. 529–543, 1986.
[52] H. Demsetz, "The cost of transacting," *Q. J. Econ.*, 1968.
[53] Grossman, Miller, and Merton, "Liquidity and market Structure," *J. Finance*, vol. 51, no. 3, 1988.
[54] H. Stoll, "Friction," *J. Finance*, vol. 60, no. 4, pp. 1479–1514, 2000.
[55] Lakonishok and Lev, "Stock Splits and Stock Dividends: Why, Who, and When," *J. Finance*, vol. 42, no. 4, pp. 913–932, 1987.
[56] Kryzanowski and Zhang, "Trading Patterns of Small and Large Traders Around Stock Split Ex-Dates," *J. Financ. Res.*, vol. XIX, no. 1, pp. 75–90, 1996.
[57] C. J. Muscarella and M. R. Vetsuypens, "Stock splits: Signaling or liquidity? The case of ADR 'solo-splits,'" *J. financ. econ.*, vol. 42, no. 1, pp. 3–26, 1996.

[58] C. M. C. Lee, "Market integration and price execution for NYSE-listed securities," *J. Finance,* vol. 48, no. 3, pp. 1009–1038, 1993.
[59] Conroy, Harris, and Benet, "The effects of stock splits on stock liquidity," *J. Econ. Financ.,* vol. XLV, no. 4, pp. 1–17, 1990.
[60] J. A. Ohlson and S. H. Penman, "Volatility increases subsequent to stock splits: An empirical aberration," *J. financ. econ.,* vol. 14, no. 2, pp. 251–266, 1985.
[61] J. A. Ohlson and S. H. Penman, "Volatility increases subsequent to stock splits: An empirical aberration," *J. financ. econ*., vol. 14, no. 2, pp. 251–266, 1985.
[62] D. A. Dubofsky, "Volatility Increases Subsequent to NYSE and AMEX Stock Splits," *J. Finance,* vol. 46, no. 1, pp. 421–431, 1991.
[63] J. J. Angel, "Tick size share prices and stock splits," *J. Finance*, 1997.
[64] Admati and Pfleiderer, "A Theory Of Intraday Patterns - Volume And Price Variability," *Rev. Financ. Stud.*, 1988.
[65] H. R.Stoll, "Inferring the components of the bid-ask spread theory and empirical tests," *J. Finance,* 1989.
[66] T. J. George, G. Kaul, and M. Nimalendran, "Estimation of the bid - ask spread and its components: a new approach," *Rev. Financ. Stud.*, vol. 4, no. 4. pp. 623–656, 1991.
[67] Chakraborty, M., "The Equity Market around the Ex-Split Date: Evidence from India," *Vikalpa,* vol. 37 no. 1. pp. 57-68, 2012.
[68] Rambhia, Rohan, Mayank Joshipura, and Nehal Joshipura, "Low risk anomaly: A new enemy of market efficiency," *IUP Journal of Financial Risk Management* vol.10, no. 3, 2013.
[69] Dash, Mihir, and Amaresh Gouda, "A Study on the Liquidity Effects of Stock Splits in Indian Stock Markets," 2007.
[70] Bajaj, Parminder, and Harman Arora, "Effect of Stock Split on the Shareholder's Wealth and Company's Profitability," *International Journal of Engineering and Management Research (IJEMR)* vol. 7. no. 1. pp. 20-27, 2017.
[71] Joshipura, Mayank, "Price and Liquidity Effects of Stock Split: An empirical evidence from Indian stock market," *NSE Research Series*, 2008.
[72] Charitou, Andreas, Nikos Vafeas, and Charis Zachariades, "Irrational investor response to stock splits in an emerging market." *The International Journal of Accounting*. vol. 40. no. 2. pp. 133-149, 2005.

Employability through networking: A way forward

Bilal Pandow, Dr. Ali Salem

Middle East College, Muscat, Oman

Abstract: The social networking is fast replacing traditional means of networking especially now it has become inevitable in the context of employment. In the present context, this study tries to unveil some of the major determinants of employability in the Middle East scenario. For the purpose, three determinants; networking; knowledge; and social capital, are identified during the group discussion and the same was tested empirically later in the study. A total of 370 questionnaires were circulated online, but only 263 responded to the study. The confirmatory factor analysis and structural equation modelling were used to perform variable modelling and recommendation were made based on the findings.

Keywords: Networking; Knowledge; Social Capital; Employability

Reference to this paper should be made as follows

1 Introduction

As the recent moves in learning and labor market strategy have stemmed in universities being positioned under aggregative pressure to produce employable graduates [1]. Also, it has implications for the environment of delivery and teaching in the higher education. Particularly there is a strain between alumni as a state subsequent in the accomplishment of a progression, and employability as a valuation of the financial worth of a student at that time [2].

Similarly, researcher have discuss three tools for improving employability in these turbulent and volatility labor markets context i.e. identity work, training and networking [3]. There have been studies conducted that relevels that the direction and self-knowledge related with self-awareness may deliver a source of inspiration, boosting, graduates to aggressively engage in job hunt in the labor market [4]. Besides, it has been argued in the literature that employability can be implanted in any academic curriculum in education without conceding core educational autonomy [5].

2 Literature review

A. Networking

On top of the classical way of networking, social or social-media networking is underpinned by a myriad of websites and applications allowing a huge number of people across the globe to interact and gather around common interests. Family members, friends, students/teachers, businessmen etc., would meet online to create and reinvigorate numerous e-social environments. [6] Social networking is becoming increasingly a key success driver in social and professional life and it is covering hundreds of different fields such as business & management, history and culture, politics etc. [7]. Many e-tools such as Facebook, Linked-In, WhatsApp, Twitter, Instagram, Snapshot etc., have been successfully facilitating the connection between billions of people worldwide.

Teachers in higher education are using these e-tools to socialize on a daily basis, so are doing their students usually called "Internet or digital generation". It would have been then very helpful if both parties were making realistically use of these social media e-tools to enhance their communication and impact positively the teaching and learning environment [8]. In fact, platforms are being used but not that much as many experts would draw the attention that "In the past five years social networking sites have drastically gained in popularity, but many users have yet to discover the true power behind effectual social networking. Surprisingly, few academics and educational administrators use free social networking sites despite the fact that online communities such as Facebook.com originated with an academic focus." [9]. Social networking could be definitely an academic cross-platform way out to get higher education students involved in a variety of a very fascinating activities for their performance. [10] An increasing number of scholars are highlighting the importance of social media e-tools in teaching and learning practices [11]. If taken into consideration and applied in a proper way, social networking should lead to enhance student's social capital as well.

B. Social Capital

Students in higher education are often encouraged to work on their social capital while studying or at least after graduating as it is never too late. The concept is based on the "photo" they would like the society to look at [12]. It tells partially the society and specifically the employers who these students/graduates really are? How are students relatively making use of their time? What type of people are they connected to? How are they selecting them? Are students socializing with interesting and invaluable people? How do they feel about themselves? [13] The side we would like to focus on here in this paper, is the students' ability to get connected with people who are close to their interest and future career. As we know social networking is a broaden world, students should be eager to network with teachers, professionals, experts etc., from the field they are planning to head to and work for. Experts are very much advising to take that direction in order to meet interesting personalities among whom students can find out the existence of very influential people [14]. The professional world is nowadays open to receive a huge influx of information, which make it therefore specifically a unique interface to meet and greet young educated people. Students can socialize online via these large variety of e-tools or even through physical meetings where we can see professionals more than happy to share their social information contacts [15]. Hence there is certainly no excuse for students to avoid getting connected in order to enhance their social image. Many HR senior managers are impressed by interviewees who both would show educational knowledge and their ability to social-network with people in the fields they would like to work in. It shows and reflects knowledge, interest, enthusiasm and motivation. These types of interviewees are unfortunately rare at the present time. What they are doing in terms of social networking with people close to their specialization is definitely adding value to their resumes [16]. It is also allowing them to leverage their recognition with respect to the society they are part of [17]. Many employers would consider the ability of social networking as a skill per.se. [18]. It is definitely part of students' investments, hoping an inevitable return along with a certain knowledge of the market.

C. Knowledge of the market

On top of social networking and getting their social capital enhanced, higher education students would also need to learn about the segment of the market they are targeting to increase their chance of employability. They could for instance have a look on the market outcomes and aligned their majors to as it will definitely help them to go the right way [19]. It then is of significance for students and/or graduates of higher education to know the nuts and bolts of the organizations and their needs they would like to work for [20]. For instance, they can find out the core competency the industry is looking for? Higher education students and/graduates should also get extra knowledge such as the size of the market of their interest as compared to the rest. They might want to know the market contribution to the GDP in percentage whether it is part of the first, the second and maybe the third sector. How the market is evolving or what are the trends over time? What are the challenges students might face? [21]. Are there journals, articles etc., about the market and what is the content focus on? Are there seminars, workshops etc., on the targeted industry and what would the upcoming outlook look like? Besides, it would not be a bad idea to inquiry about the taxonomy of companies that are operating in the targeted market. For instance, companies could be classified by the type of products they are making and/or selling and their size in terms of numbers of employees, turnover, the amount of value added created, share of the market etc. [22]. Higher education students should consider knowing, the "kings and/or the queens" of the market as well, at least by their names. They need to find out and try to get acquainted with key persons in as many companies as they could, and have an idea on what their days look like. Here, trainings, internships and/or volunteering experiences would be helpful [23]. Students are finally urged to handle live projects and help companies to solve problems within a win-win innovative and entrepreneurial setting [24].

D. Employability

As the concept of employability is now increasingly embedded in higher education's institutions [25], social networking, although helpful [26], would not lead most of the time and directly to that goal. The argument raised here through this paper is that social networking is first of significance for higher education students or graduates' social capital and knowledge of the market. These two intervening variables should ultimately show their positive implications for employability. On the one hand, potential employers would definitely look for candidates who would successfully pass their writing test (if any) showing their higher education experience aligned with what York was stating in these terms "Although a student's experience of higher education cannot guarantee a 'graduate level job', the nature of that experience influences the chances of success" [27]. On the other hand, they are expected candidates to be more articulate during their interviews and that would not happen without a bridge between quality of learning and employability [28]. Employers are also looking for impressive candidates during interviews where they would expect some questions on the organizations they intend to work for [29]. Also, researcher have observed and recommended the business trainings during the students are studying. While the focus should be given on providing more skill based learnings in the educational institution in Oman so as to increase their employability [30]. Now, what if a candidate was even ready to provide a full description of the entire segment of the market of his interest? What if a candidate is able to demonstrate his enhanced social capital? Reinforcing the idea that the candidate is interested to work for the targeted organization will certainly make a difference [31]. Higher education students/graduates should reach the level of being able to sell themselves in a very professional manner. Social image and knowledge of the market are key factors here and a good number of employers would not hesitate to hire candidates on the spot!

3 Research Gap

There have been plethora of studies conducted on employability in the past and also evaluated the various factors affecting it. However, most of these studies followed a similar set of constructs which were identified by previous studies. Most of the researchers have used the replica set of constructs and scales for different regions and countries without conducting preliminary interviews. In Middle East context, the number of studies conducted is limited, and most of them have used similar set of variables and the scales for their research. Also, the other vital gap that was found is that many of the researches have used regression analysis to access the impact on the employability. This has limited the study as the extrapolation could be premeditated as a summated power of the forecasters rather than calculating the individual forecast power of every single forecaster.

A. Contribution of the Study

The present study is unlike the rest of the previous researches conducted in the similar sphere as it ascertains a list of the factors of employability with the help of networking using a focal group practice. The study safeguarded to reduce the researchers' prejudice to choose among set of random theories found appropriate and then to assess them. The focus group discussion also validated the point that factors that are significant to Middle East youth may not be as significant to people of other regions as has been stated in the literature analysis. This study also tries to highlight precisely the factors affecting the employability and its linkage with networking, social capital and knowledge.

B. Research Objectives

The main aim of the study is to ascertain the major factors that affect the employability. The other objectives are mentioned below:

1. To examine the relationship between networking and employability.
2. To examine how social capital affects employability.
3. To assess the impacts of knowledge on employability.

4. To examine if the networking has an impact on social capital and knowledge.

4 Methodology

The hypothetical deductive approach through a structural equation models is used in the study, in a population of 12,551 of which 263 respondents from various departments responded to our questionnaire which was sent online using google forms. The conceptual model designed by the researchers is having one independent variable (Networking), two mediating variables (Social capital and Knowledge of the market) and finally one dependent variable (Employability).

A. Research Design

The study was conducted to investigate the significant factors affecting employability. After focus group discussions, the present study scaled down to only three factors. After conducting a literature review, the study found that different studies have reinforced each of the anticipated factors as a contributor to employability of the youth in Sultanate of Oman. The next step was to get a realistic confirmation of the same. The study used the measureable approach with survey approach to fulfil this objective. To administer the survey, the study used questionnaire technique and an instrument (Tab. 2) was developed in the course of the study while keeping an eye on the available literature.

The instrument contains 14 items representing Networking (3 Items), Knowledge (3 Items), Social capital (4 Items), and Employability (4 Items).

B. Sample Design

The distribution of respondents is provided in below mentioned (Tab. 1).

TABLE 1: DISTRIBUTION

Variables	Particulars	N	%age
Gender	Male	114	43
	Female	149	57
Networking	face-to-face	17	6.5
	online	244	93.5

Distribution of respondents from the survey

C. Data Collection

The data was collected through a self-administered questionnaire. In total 370 questionnaires were mailed to the respondents out of which only 283 were received back. After deleting responses with more than three missing answers and removing outliers, only 263 remained. The respondents were distributed on various demographic factors like gender, age, education and occupation.

Besides, the respondents were asked where they usually attend social networking. Almost, 73% respondent mentioned educational institutions/colleges, followed by 11%

clubs, 9% mentioned NGOs, while rest prefer private firms, government offices and others.

D. Sampling Technique

In this paper we have used judgmental sampling as the questionnaire was mailed to various people online. However, no pre-set condition was laid to select the respondents. The questionnaire was also shared on social media sites like G+ and Facebook. Also, we shared it on various android apps like: WhatsApp, imo and viber.

TABLE 2: CONSTRUCT AND MEASURES

	Employability through Networking
Networking N1	I see networking as a way to socialize
N2	I see networking as a way to express myself
N3	I see networking as means to be recognized
Knowledge K1	Networking is getting me extra jobs skills
K2	Networking is improving my interpersonal skills
K3	Networking is helpful to know the segment of the market I would like to work in
Socialcapital S1	Networking has allowed me to interact with influential people
S2	Networking has allowed me entry into the professional world
S3	Networking is adding value to my resume
S4	Networking has given me improved recognition
Employability E1	I am getting positive response from potential employers
E2	Employers see me as a knowledgeable individual
E3	I am being viewed as a worthy candidate
E4	I have received some job offers

5 Results and Discussions

The majority of the respondent for this study to the extent of 57 percent are females, also most of the respondents are engaged with the online social networking, and the same can be observed from the (Tab.1). The study has been conducted in the Muscat, capital of Sultanate of Oman, in two leading higher educational institutions that draws students from almost all the regions of the country. For this study we have used the Smart-PLS software for the data analysis. Beside, based on the focal group discussion the researchers have set the questionnaire and below mentioned hypothesis.

A. Hypothesis Development

- H1: Networking is highly and positively related to the social capital
- H2: Networking is highly and positively related to knowledge of the market
- H3: Knowledge of the market is positively impacting and moderating the employability
- H4: Social capital is positively impacting and moderating the employability

B. Construct Reliability and Validity (CRV)

The Cronbach's alpha analyses the indicators for latent variables exhibit convergent strength, hence present reliability. By convention, a good scale should possess greater or equal to 0.80 as is the case here having value of 0.867. Also, Average Variance Extracted (AVE) reflects the average communality for each latent element in a reflective model. In an adequate model, AVE should be greater than 0.5 [32] and in the present study we have 0.717. In a model satisfactory for exploratory purposes, composite reliabilities should be greater than 0.6 and the composite reliability 0.910 as is mentioned in (Tab.3) which is the case for this study which is considered worthy for confirmatory research.

TABLE 3: CONSTRUCT RELIABILITY AND VALIDITY

	Cronbach's Alpha	***Composite Reliability***	***Average Variance Extracted***
Employability	0.867	0.910	0.717

The path coefficients are standardized path weights consequently vary from -1 to +1. The weights nearby 1 reveals the strongest paths while the one near to 0 reflects the weakest path. In the above, the path coefficients of knowledge, social capital are having strongest path followed by networking. While the moderation effect 2 is also having a significant impact on the employability, however, the moderation effect 1 is having a negative impact on the employability also can be seen in (Tab.4).

TABLE 4: PATH COEFFICIENTS

	Employability	***Knowledge***	***Social Capital***
Networking	0.081	0.705	0.703
Knowledge	0.143	-	-
Social Capital	0.669	-	-
Moderating Effect 1	-0.057	-	-
Moderating Effect 2	0.154	-	-

The path coefficients as represented in the model can be observed in the below mentioned (Fig.1).

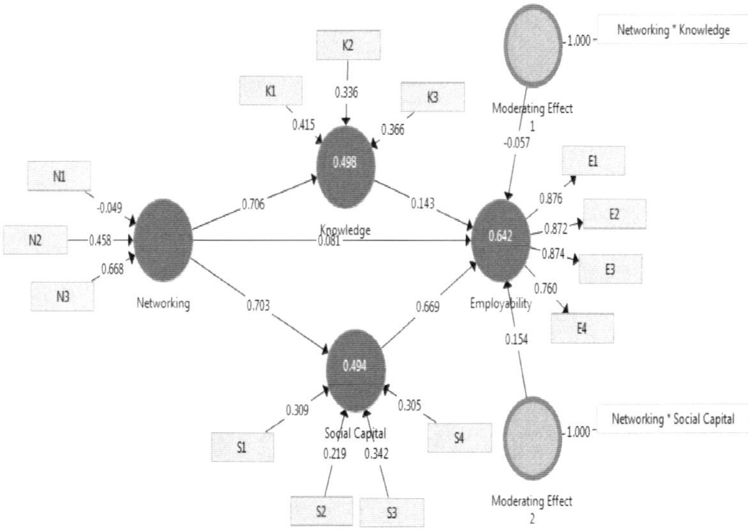

Fig. 1: Partial least squares model

There is an indirect effect of the networking through social capital which is having an effect on the employability to the tune of 0.47 with p-value of 0.000 showing the high degree of significance, while the networking through knowledge and in turn on employability is not having that strong effect as can be seen in (Tab.5). This makes us to accept the H4 hypothesis and at the same time reject the H3 hypothesis.

TABLE 5: INDIRECT EFFECT

	Specific Indirect Effects	*P Values*
Networking -> Knowledge -> Employability	0.101	0.049
Networking -> Social Capital -> Employability	0.470	0.000

The measurement loadings are the regular path weights linking the factors to the pointer variables. The loadings usually vary from 0 to 1. The larger loadings mean stronger and reliable the dimension model. By convention, for a well-fitting model, path loadings should be above 0.70 and in the case of this study, it can be observed that the loadings of all the variables are above 0.70 except N1 as can be seen from (Tab.6).

TABLE 6: OUTER LOADINGS

	Employability	*Knowledge*	*Networking*	*Social Capital*
E1	0.876			
E2	0.872			
E3	0.874			
E4	0.760			

	Employability	Knowledge	Networking	Social Capital
K1		0.906		
K2		0.891		
K3		0.888		
N1			0.678	
N2			0.876	
N3			0.947	
S1				0.826
S2				0.825
S3				0.893
S4				0.849

The importance performance map analysis of the model reveals that the path coefficient of the networking and knowledge is at 0.696 which is statistically significant having an LV performance indicator of networking at 65.054 and knowledge at 72.161. While as the path coefficient of networking and social capital is at 0.617 which also is statistically significant having an LV performance indicator of social capital at 70.611 and can been seen in the (Fig.2). Thus accepting the H1 and H2 hypothesis. This is also substantiated by the p-values mentioned in the (Tab.7).

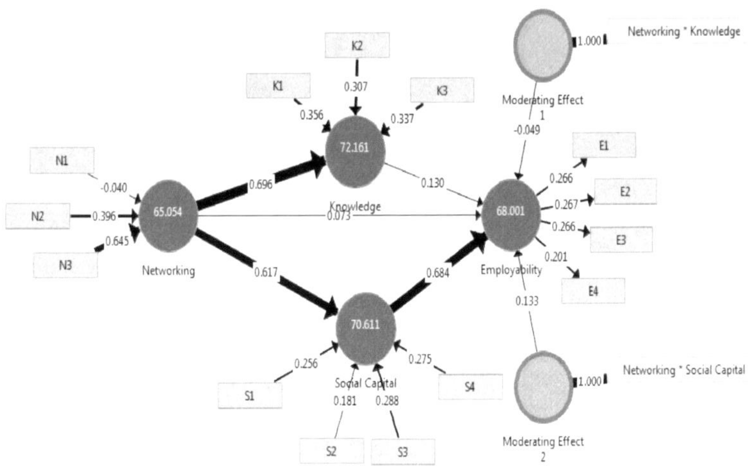

Fig. 2: Importance Performance Map Analysis model

TABLE 7: PATH COEFFICIENTS

	Mean	SD	T Stat	P Values
Knowledge -> Employability	0.156	0.076	2.028	0.043
Networking -> Employability	0.084	0.083	1.047	0.296
Networking -> Knowledge	0.711	0.038	18.538	0.000
Networking -> Social Capital	0.710	0.038	18.653	0.000
Social Capital -> Employability	0.720	0.075	9.602	0.000
Moderation 1	-0.042	0.084	0.722	0.471
Moderation 2	0.142	0.082	2.000	0.046

As can been seen from the (Tab.8) the R square for the employment latent variable is 0.735 which is good and significant and depicts about the reliability of the model used in this paper.

TABLE 8: INDIRECT EFFECT

	R Square	R Square Adjusted
Employability	0.735	0.729
Knowledge	0.498	0.496
Social Capital	0.494	0.492

6 Conclusion

The primary goal of the study was to highlight the major factors that affect the employability of the youth of Sultanate of Oman. From a diverse set of predictors that were thought to be critical in the development of employability, only three were deliberated to be of significance. The study reveals that networking does play a positive and significant role in the knowledge and building the social capital of the people. Besides, social capital does have a positive and significant impact mediating the two variables of networking and employability. However, the study also found that there is no significant impact of the knowledge variable in mediating the two variables of networking and employability. Hence based on these findings we recommend the educations institutions especially the higher education establishments to encourage students for more social networking by organizing guest lecturers, taking students for industrial tours, making internships mandatory in their curriculum and promoting students to undertake live project. The implementation of all these recommendation would surely fetch the results and would increase the employability of the students especially in the Middle East region where the governments are working hard to see their youth getting employed.

7 Future directions and limitations

This study is beneficial in the area of higher education. However, it has its limitations. The first and far most important limitation of the study is that of the sample size. The study that has a population in tens of hundreds needs a substantial sample size to validate the outcomes. However, due to time and financial restrictions, it was not promising to go beyond a convinced number for the present study. This is one of the points that researchers occupied in this arena shall consider for future research. Besides, the sampling technique that slightly restricts the study is one such area where future researchers can improve upon using probability procedures. Also, another limitation is that these findings are based on the expressed attitudes of the respondents who are exposed to usual mistakes characteristic in human nature.

References

[1] Bridgstock, R.,. "The graduate attributes we've overlooked: Enhancing graduate employability through career management skills,". Higher Education Research & Development, vol. 28. no. 1. pp.31-44. 2009.

[2] Glover, D., Law, S. and Youngman, A.,. "Graduateness and Employability: student perceptions of the personal outcomes of university education,". Research in post-compulsory education, vol. 7. no.3., pp.293-306. 2002.

[3] Smith, V.,. "Enhancing employability: Human, cultural, and social capital in an era of turbulent unpredictability,". Human Relations, vol.63. no.2. pp.279-300. 2010.

[4] McArdle, S., Waters, L., Briscoe, J.P. and Hall, D.T.T. "Employability during unemployment: Adaptability, career identity and human and social capital,". Journal of vocational behavior, vol.71. no.2. pp.247-264. 2007.

[5] Knight, P.T. and Yorke, M. "Employability through the curriculum,". Tertiary Education & Management, vol.8. no.4. pp.261-276. 2002.

[6] Kadushin, C. "Understanding Social Networks: Theories, Concepts, and Findings,". Oxford: Oxford University Press. 2012.

[7] Neher, K. "Social Media Field Guide: Discover the Strategies, Tactics and Tools for Successful Social Media Marketing,". Boot Camp Digital Publishing. 2010.

[8] Junco, R. "Engaging Students Through Social Media: Evidence-Based Practices for Use in Student Affairs,". 1st Edn. USA: Jossey-Bass. 2014.

[9] Michael D. King. "Why Higer Ed and Business Need to Work Together, Global Education Industry,". Available from https://hbr.org/2015/07/why-higher-ed-and-business-need-to-work-together. June 2015.

[10] Al-Tarawneh, H. "The Influence of Social Networks on Students' Performance,". vol.5.pp. 200-205. Available from https://pdfs.semanticscholar.org/a3a2/492eb9dc2cfcf36848261ae2df5948d55f2e.pdf. 2014.

[11] Rosmala, D. "Study of Social Networking Usage in Higher Education,". Elsevier, vol. 67. pp. 156-166. 2012.

[12] Rodriguez Mosquera, P., Uskul, A., and Cross, S. "The centrality of social image in social psychology,". European Journal of Social Psychology. vol 41. pp. 403-410. 2011.

[13] Canevelo, A., and Crocker, J. "Interpersonal goals, others' regard for the self, and self-esteem: The paradoxical consequences of self-image and compassionate goals,". European Journal of Social Psychology. vol 41, pp. 422-434. 2011.

[14] O'Hearn, A. "How to Meet the Right People to Get You Your Dream Job,". Retrieved online February 10, 2015, from https://www.inc.com/aaron-ohearn/why-you-haven-t-met-the-right-people-to-get-your-dream-job.html

[15] Fishbein, M. "How to Build an Awesome Professional Network,". Createspace Independent Pub. 2014.

[16] Clark, A. "Ways You Can Show Networking Strength on a Resume and Cover Letter,". Retrieved online July 3, 2014, from https://www.business2community.com/human-resources/ways-can-show-networking-strength-resume-cover-letter-0934768

[17] Simon, B., & Renger, D. "Social recognition as an equal: The role of equality-based respect in group life,". European Journal of Social Psychology. vol 41, pp. 501-507. 2011.

[18] Harris, B. "Get Hired: Learn Six Employer Secrets That Can Improve Your Cover Letter, Resume, Networking Skills, and Job Interview Results to Help You Get Hired,". Createspace. 2011.

[19] Baker, R., Bettinger, E., Jacob, B., and Marinescu, I. "The Effect of Labor Market Information on Community College Students' Major Choice,". National Bureau of Economic Research. Michigan. 2017.

[20] Thune T., and Støren L.A.. "Study and Labour Market Effects of Graduate Students,". Interaction with Work Organisations During Education: A Cohort Study. Education + Training. vol. 57. No.7, pp. 702-722. 2015.

[21] Gill A., Lashine S. "Business Education: A Strategic Market-Oriented Focus,". International Journal of Educational Management. vol. 17. no.5, pp. 188-194. 2003.

[22] Craft, D. "Company Profiles for Students. The Gale Group,". 1999

[23] Steven Kelly, J and Stevenson, K. "Cases by Outstanding Marketing Students: DePaul University 2016-2017,". Midwest Marketing Education Foundation. 2017

[24] Babinska, D., Witczak, O. "'Developing students' knowledge, skills and social competences in international marketing research – cooperation of academia and business,". Journal of Economics and Management. vol.26. no.4. 2016.

[25] Norton, S. "Embedding Employability in Higher Education for Student Success,". Higher Education Academy. 2016.

[26] Adler, L. "New Survey Reveals 85% of All Jobs are Filled Via Networking,". Performance-based Hiring Learning Systems, USA. Retrieved online December 21,

2017, from https://www.linkedin.com/pulse/new-survey-reveals-85-all-jobs-filled-via-networking-lou-adler. 2016.

[27] Yorke, M. "Encouraging the Development of Employability,". Briefing on Employability. 2010.

[28] Knight, P. and Yorke, M. "Assessment, Learning and Employability,". London: Society for Research into Higher Education and Open University Press. 2003.

[29] Raskin, Don. "The Dirty Little Secrets of Getting Your Dream Job,". Simon and Schuster. 2016.

[30] Pandow, B. and Salem, A. "Evaluating readiness of youth to start business: A study in Oman,". 2017.

[31] Mayhew, R. "What Factors Affect Expectations in an Interview?," Retrieved online January 4, 2017, from http://work.chron.com/factors-affect-expectations-interview-18029.html. 2017.

[32] Chin, Wynne W. "The partial least squares approach to structural equation modeling,". Modern methods for business research vol.295. no. 2. pp.295-336. 1998.

Challenges of Using MIMO Channel Technology in 5G wireless Communication Systems

Dr.Intisar Al-Mejibli
Informatics Systems Management Dept.,
University Of Information Technology and Communications,Baghdad,Iraq
Dr.intisar.almejibli@gmail.com

Dr.Salah Al-Majeed ,Senior member IEEE,Srilatha Pamuri
Systems Engineering Dept.,
Military Technological College,
Muscat,Oman
Salah.almajeed@mtc.edu.om

Abstract—**Multiple input multiple output (MIMO) systems, which are suitable for mmWave communications, facilitate building large arrays of communication channel with Reasonable form factors indoor and outdoor communications. In both, indoors and outdoors communications a reasonable coverage can be achieved by the high gains of array. Deploying MIMO systems in frequency-division duplex mode faces the issue of high feedback overhead of channel state information (CSI) to the transmitter. This issue is attributed to the used multiplexing algorithm. From other hand, it is expected that by 2020, all new 5G technologies will be in operation. Hence, it is crucial to investigate and analyze using the MIMO system with orthogonal frequency-division multiplexing OFDM modulation in 5G technology. This research sheds light on the potential challenges of implementing MIMO-OFDM in 5G technology. It proposed four models with different number of transmitter and receiver antenna elements in the array. All proposed model used 2 elements per row in transmitter and receiver antenna. The four models are implemented in two cases Line of sight LOS and Non Line of Sight NLOS. The channel rank and condition number of channel matrix are regarded as significant and accurate mathematical indicators which can be used to evaluation MIMO channel technology. The condition number is calculated from the obtained results of four models with rank of channel metrics.**

Keywords—***5G; Channel Coefficient*** $\mathbf{h_{(m,k)}(f)}$ ***; MIMO ; NLOS; OFDM; SISO.***

I. Introduction

The demand on higher data rates, services with better quality and fully mobile and connected wireless network is increased. 5G technology satisfies the previous demands as it is anticipated that 5G wireless networks will achieve 10 times higher than 4G in spectral efficiency and energy efficiency. That required significant changes in the layers design of 5G communication systems. Multiple input multiple output (MIMO) technologies is one of candidate for the physical layer design of 5G networks. MIMO based on transmitting different signals over multiple antennas to gain capacity. During the mid-1990s, Research studies showed that, though combining MIMO with popular air interfaces for instance time-division multiple access (TDMA) and code-division multiple access (CDMA) improves communications quality, using MIMO and OFDM is most effective at higher data rates [1]. Using OFDM modulation with MIMO will provide more reliable communications at high speeds because OFDM divides a radio channel into a large number of sub channels that are closely spaced [2].

Soon, 5G will impact all people lives in multiple areas such as education, healthcare, transportation and banking. 5G technology will developed these industries technically, in the same time, the network specialists must work hard to make this technology commercially viable and to overcome any challenges may appear. One of the significant aspect is ensuring the cost of using the 5G mobile network reasonable regardless the implemented industries. As the vast range of 5G industries, specialists will need to identify and manage several new commercial arrangements and cost structures [3]. Moreover, great effort must be made by scientists and specialists to study, analyze implementing 5G technology using current wireless communication techniques and technologies.

This research investigates and evaluates using MIMO-OFDM in 5G wireless communication system by implementing four models, each with distinct parameters. These models calculate and compare the condition number and rank of channel matrix from the resulting models files.

The rest of this paper is organized as follows: section two presents the 5G technology, and MIMO channel is described in section three. Evaluate MIMO channel is explained in section four. Section five introduces the simulated models. Conclusion is described in section six.

II. 5G Technology

Although, the current status of the 5G technology is its early development stages, the 5G technology and its services have attracted the attention of specialists and researchers. Currently, the researches that study, analyze and investigate 5G technology from different aspects are still few and needs more effort from the researchers to enrich this field. Generally, advanced technologies could be used to become part of a new 5G system because of that several companies are looking into these technologies. Additionally, 5G research units that focus on developing the 5G technologies, have established in many universities.

There are a common challenges that faced 5G technology such as research methodology, lack of infrastructure, and cost. From the other hand, standardization is regarded as the main challenging in 5G technology as it it's critical to ensure universal interoperability of this new technology [3]. From the point of view commentator state: 5G is ubiquitous access to services with high & low data rate.

III. MIMO Channel

MIMO technologies are the key variables of state-of-the-art mobile radio systems and are essential to accomplish extremely ambitious objectives of capacity such as giving stable data rates [4].

MIMO channel is the channel that use both multiple transmit and multiple receive antennas. MIMO channel provides extra spatial dimension for communication, results in a degree-of-freedom gain via beam forming and increase reliability. These additional degrees of freedom can be exploited to spatially multiplexing many of data streams onto the MIMO channel which results in increasing the capacity. The n transmit and receive antennas in such MIMO channel capacity is proportional to n [5]. A basic 2x2 spatial multiplexing is showed in Fig. 1.

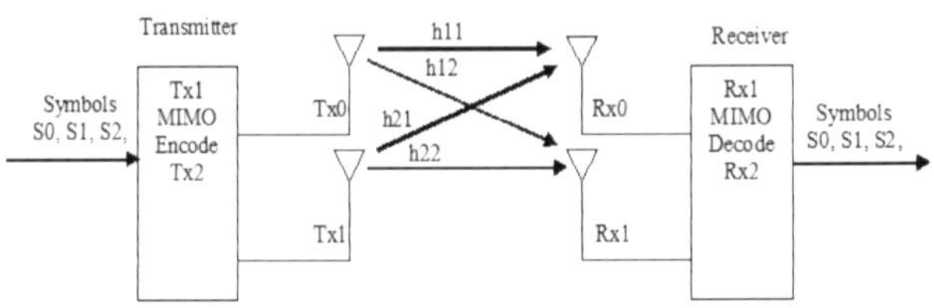

Fig. 1. MIMO configuration of 2x2 spatial multiplexing with the associated four complex channel coefficients.

A continuous evaluation of complex matrices is required to evaluate the status of a MIMO channel. The channel rank and condition number are viewed as generally straightforward mathematical indicators that can be used to measure the quality of MIMO channel [5].

The maximum channel capacity is calculated by using the bandwidth B and the effective signal-to-noise ratio (SNR). In accordance with the Shannon-Hartley theorem for single input single output (SISO), the maximum channel capacity C is calculated as expressed in formula 1 [6] :

$$C = B * \mathrm{Log}_2 \left(1 + \frac{P}{N}\right) \qquad (1)$$

where P is the received signal power and N is the noise power. Achieving the targeted peak data rates essentially depend on the available channel bandwidth B because B has a directly proportionate effect on the channel capacity.

In wireless communication, the signal-to-noise ratio is a function of the physical distance between the receiver (such as a smart phone) and the base station, so it is location-dependent value. The performance or transmission quality is mainly affected by SNR which is decisive for it and therefore for the actual data throughput in a wireless cell [7].

These two channel attributes can be improved by using designed MIMO system and without increasing the bandwidth or transmit signal power. Diversity of transmit and/or receive antenna can improve the channel performance. Increasing the channel capacity can be achieved by apply spatial multiplexing. Consequently, apply spatial multiplexing alone results a significant increase in the channel capacity because it offers the possibility of multilayer data signals transmission. Calculating the MIMO channel capacity will become very complicated process compared to the aforementioned Shannon-Hartley theorem for (SISO). However, calculating the SISO channel capacity as M-times by Shannon can be used to estimate the greatest possible MIMO channel capacity as M-layer data transmission [8]. Better channel conditions and higher SNR than an equivalent SISO system are required for successful spatial multiplexing.

IV. Evaluate MIMO Channel

Evaluate the MIMO characteristics of a mobile radio channel requires an appropriate mathematical model that includes set of linear equations [7]. Transmit and receive signal vector form are used to represent the transmitter and receiver respectively. The real transmission feature, or the present channel state, is summarized in a matrix which is useful in assessing the MIMO channel [7].

The condition number is characterized as the proportion of the largest to smallest singular value in the singular value decomposition of a matrix. Additionally, in the context of wireless communications, it is a metric to characterize the quality of MIMO channels [9].

A well-conditioned system of linear equations has small condition number value (e.g., below 20 dB) where small errors in the matrix coefficients result in small errors in the solution. A well-conditioned system indicates that each sub-channel usually has a distinct spatial direction. An ill-conditioned system of linear equations has high condition number value (e.g., over 20 dB) where small errors in the coefficients may have a large detrimental effect on the solution [10].

The following formula represents how the condition number K(H) of matrix H is calculated:

$$K(H) = \frac{\sigma_{max}}{\sigma_{min}} \geq 1 \quad (2)$$

where σ_{max} represents the largest singular value and σ_{min} represents the smallest singular value in matrix.

The rank of a matrix is the dimension of the vector space generated (or spanned) by its columns (or rows), so it represents the number of singular values not equal to zero. Thus the rank of the channel matrix is an indicator of how many data streams can be spatially multiplexed over the channel in the context of MIMO communications [6][9].

V. SIMULATED MODELS

NYUSIM simulator is used to simulate millimeter-wave channel for the four proposed models. It provides precise measurements in terms of time and space. These measurements includes rendering of actual channel impulse responses and realistic signal levels [9].

This simulator provides several output files that includes information of each resolvable multipath component. The conducted measurements from simulator can be used to generate the MIMO channel coefficient for an OFDM sub-carrier. The following equation (3) [11] is used to compute the channel coefficient

$$h_{m,k}(f) = \sum_p (a_{m,k,p} e^{j\emptyset_{m,k,p}} e^{-j2\pi f T_{m,k,p}} \times e^{-j2\pi f d_T \, \mathrm{msin}(\emptyset_{m,k,p})} \times e^{-j2\pi d_R \, \mathrm{ksin}(\varphi_{m,k,p})}) \quad (3)$$

Where the MIMO channel coefficient is represented by $h_{m,k}(f)$. The mth transmit antenna and the kth receive antenna represent the range of MIMO channel coefficient for the sub-carrier f. The parameters in equation 3 are explained in Table 1.

Table 1. The Meaning of Equation 3 Parameters

Parameter	Meaning
p^{th}	resolvable multipath component
α	the amplitude of the channel gain
ϕ	multipath component phase
φ	departure azimuth angle and arrival angle
d_T	antenna element spacing of transmitter
d_R	antenna element spacing of receiver

All these parameters can be extracted from the NYUSIM simulator files after executing the model. In a MIMO-OFDM system, there exists an $N_t * N_r$ channel matrix that named as H. Where N_t represents the number of transmitter (TX) antenna elements and Nr represents the number of receiver (RX) antenna elements. The elements of H matrix are $h_{m,k}(f)$, where $m = 1,\ldots,N_t$ and $k = 1,\ldots,Nr$. Thus, it is possible to obtain the condition number of H.

The simulated model based on using the aforementioned approach, where table 2 shows the used values as input parameters in the NYUSIM GUI.

Four simulated model are implemented which are different in number of N_t and N_r values. The 1st,2nd,3rd,and 4th simulated models use different sizes of communication channel arrays described by these values: (1*2) =2, (2*2) =4, (3*2) =6 and (4*2)= 8, where number of columns takes the value of (1,2,3, and 4) and the antenna elements per row is fixed (2) in transmitter and receiver.

The simulated model takes into consideration the weather factors of Baghdad city in July. So the used values in simulated model are as following: temperature =43°, rainfall= 0mm/hr, pressure= 998.5 mbar, and humidity= 12% [12]. The used weather values represent the mean values in July at Baghdad.

Table 2. The values of input parameters on the NYUSIM GUI

Parameter	Value
Frequency	28 GHz
RF bandwidth	800 MHz
environment	outdoor in urban microcell (UMi)
Environment	Line of sight (LOS) and Non Line of sight (NLOS)
Lower Bound of T-R & R-T Separation Distance	100 meter
TX Power	30 dBm
Number of RX Locations	100 locations
TX Antenna Azimuth and Elevation HPBW	10° Celsius
RX Antenna Azimuth and Elevation HPBW	10° Celsius
Number of TX and RX Antenna Elements Per Row Wt and Wr	2,2 respectively
TX and RX antenna array type	Uniform Linear Array (ULA)
TX and RX Antenna Spacing	0.5 wavelength

The simulated models assign the frequency interval between adjacent sub-carriers to 500 kHz, as this value is corresponds to 800 MHz/500 kHz and equal to1600 sub-carriers. Each simulated model is performed 100 time (because the number of RX locations is set to 100), results in emulating 100 random MIMO channel realizations. Empirical cumulative distribution function (CDF) used to compute the probability measure which is implicitly specified by the frequency counts in the obtained condition numbers [13].

Figure 2, 3, 4 and 5 illustrates the empirical CDF of the condition number of channel matrices when Nt and Nr values =2, 4, 6, 8 respectively.

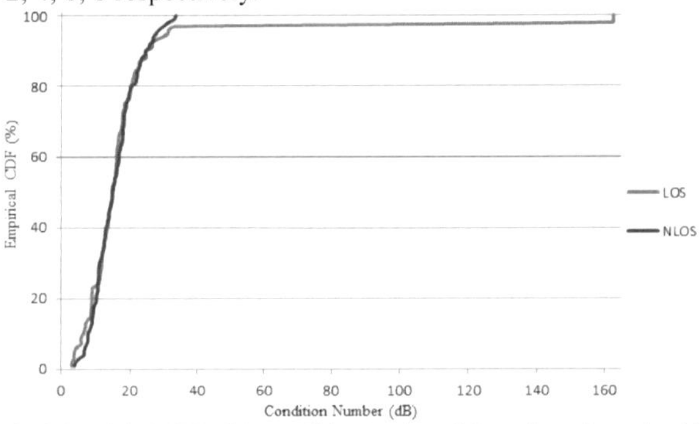

Fig. 2. Empirical CDF of the condition number of channel matrices when Nt =2 and Nr =2

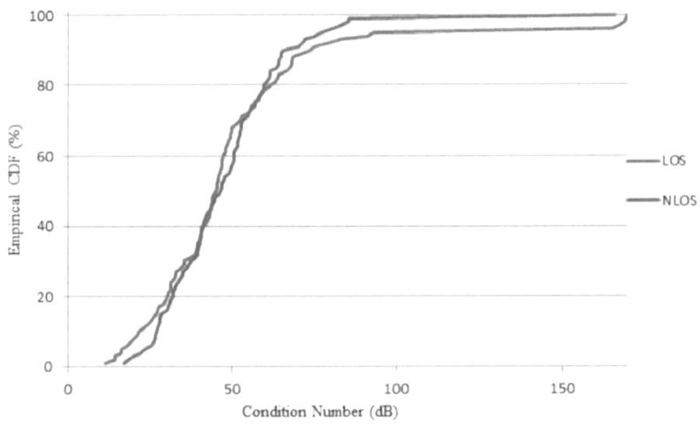

Fig. 3. Empirical CDF of the condition number of channel matrices when Nt =4 and Nr =4

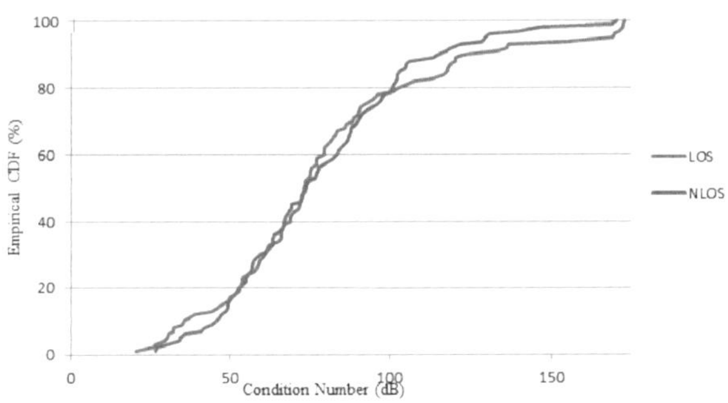

Fig. 4. Empirical CDF of the condition number of channel matrices when Nt =6 and Nr =6

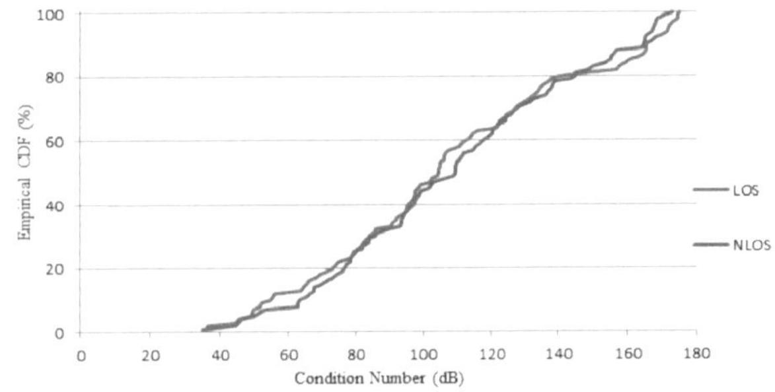

Fig. 5. Empirical CDF of the condition number of channel matrices when Nt =8 and Nr =8

Physically, good orthogonality can be achieved when condition number value is small (e.g., below 20 dB) of different sub-channels. Usually, a distinct spatial direction is specified for each sub-channel, additionally the channel gains can be compared in different spatial directions.
It is apparent from Fig.2, 3, 4, and 5 that when the Nt and Nr increased the condition number is increased too which results in bad orthogonality.
The differences in condition numbers of the individual OFDM sub-carriers between the 4 matrices in the case of LOS and NLOS are described in table 3.

Table 3. The differences between Matrices

Matrices Size	Difference
1*2	4.7 dB
2*2	6.2 dB
2*3	5.41 dB
2*4	3.57 dB

These differences are attributed to the fact of matrix rank deficient as the results shows that the probability of achieving full rank is decreased as the number of Nt and Nr increased. For example in LOS case, the average median value of the condition number is 19 dB when both TX and RX antenna elements equal to 2. This number is increased when TX and RX antenna elements increased. Consequently, the average median value of the condition number is: 50 dB, 79 dB, 107 dB when the numbers of TX and RX antenna elements are 4, 6, and 8 respectively. As a result, the spatial multiplexing technique can be used effectively to send the spatial streams simultaneously when TX and RX antenna elements less than or equal to 2.

Table 4, shows the average differences between LOS and NLOS. There is a slightly difference in two cases.

Table 4. The Average differences between LOS and NLOS

Matrices	LOS Difference	NLOS Difference
Between 1*2 and 2*2 matrices	31 dB	32 dB
Between 2*2 and 2*3 matrices	29 dB	31 dB
Between 2*3 and 2*4 matrices	27 dB	30 dB

VI. Conclusion

Using Multiple-input, multiple-output with orthogonal frequency-division multiplexing (MIMO-OFDM) is regarded as dominant air interface for 5G wireless communication systems. MIMO facilitates transmitting different signals over multiple antennas to multiply capacity. Orthogonal frequency-division multiplexing (OFDM) provides more reliable communications at high speeds by dividing a radio channel into a large number of sub channels which are closely spaced. The results of combining the technology of MIMO and OFDM is expanding the capacity and achieving more reliable communications at high speeds. MIMO-OFDM is the base for many advanced wireless local area network and network standards of mobile broadband. This combination delivers the highest data throughput and capacity by achieving the greatest spectral efficiency.

It is expected that 5G wireless networks will be deployed in next few years to satisfy the increased demand on more efficient communication system in terms of maximal data rates, service in better quality, fully mobile and connected wireless networks. Accordingly, it is significant to study deployment of MIMO with OFDM in 5G wireless networks. This research proposed four different models to study and analyze implementing MIMI and OFDM technologies in 5G. It takes into its consideration the LOS and NLOS environments. Two mathematical indicators are used to measure the quality of MIMO channels, which are condition number and rank of the matrix. The results show that the relation between the condition number and number of TX and RX antenna elements is positive relation, while the achieving full ranking matrix and number of TX and RX antenna elements are related with contrary relationship.

References

[1] Ms. Komalpokale, Prof.S. Ahirwar, and Prof. G. Ramalakshmi, "VLSI Implementation of MIMO", International Journal of Computer Application (2250-1797) Volume 5– No. 6, October 2015.

[2] Long Bao Le, Vincent Lau, Eduard Jorswieck, Ngoc-Dung Dao, Afshin Haghighat, DongInKim, Tho Le-Ngoc, "Enabling 5G mobile wireless technologies", Sipringer Leet al. EURASIP Journal on Wireless Communications and Networking, (2015) 2015:218, DOI 10.1186/s13638-015-0452-9.

[3] Ilia Abramov, product director and security expert, Xura, "5G Rising: Changes and Challenges in the Next-Generation Network", 2016,.

[4] Stefan Schindler, Heinz Mellein, "Assessing a MIMO Channel", Rohde & Schwarz, 2011

[5] David Tse and Pramod Viswanath, "Fundamentals of Wireless Communication", Cambridge University Press, 2005.

[6] Olivier Rioul, and José Carlos Magossi, "On Shannon's Formula and Hartley's Rule: Beyond the Mathematical Coincidence", journal of Entropy 2014, 16, 4892-4910; doi:10.3390/e16094892.

[7] Rohde & Schwarz: "Introduction to MIMO", Application Note 1MA142, 2009

[8] Foschini, Gans: "On Limits of Wireless Communications in a Fading Environment when Using Multiple Antennas", Wireless Personal Communications 6, 311-335, 1998.

[9] New York University and NYU WIRELESS, "NYUSIM User Manual", New York USA, 2017

[10] Keysight Technologies, "MIMO Performance and Condition Number in LTE Test", USA, 2014

[11] A. Adhikary, E. A. Safadi, M. K. Samimi, R. Wang, G. Caire, T. S. Rappaport, and A. F. Molisch, "Joint spatial division and multiplexing for mm-wave channels", IEEE Journal on Selected Areas in Communications, vol. 32, no. 6, pp. 1239–1255, 2014.

[12] Baghdad International (sda) Monthly Climate Average, Baghdad.

[13] Rui Castro, "Introduction and the Empirical CDF", 2013, http://www.win.tue.nl/~rmcastro/AppStat2013/files/lecture1.pdf

Smart Car

Haseena Bibi Zulfiqar | Ali Alhumairi | GUTECH(German University Of Technology in Oman)

Problem / Question

Naturally, robot cars can't interact the same way humans can, which's problematic.

Better software & Communication.

Greater Communication.

Hypothesis

- Before starting the project, My hypothesis was to come up with smart scenario for self- driving car. Even for self-driving car, I though what If I put a button to convert a self-driving car into self-

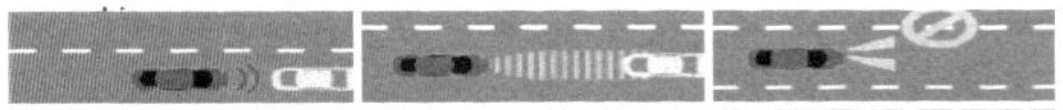

Project Overview

- My Smart car is about Self-Driving car and Self-Parking car on a Arduino Mega platform.

- Self-Driving car: For Self-Driving car I'm using Ultra-sonic to detect or sense its environment and navigating without human
- input.
- Self-Parking car: Self-parking car is an
- autonomous car-maneuvering system that moves a car into a parking spot to perform parallel, perpendicular or angle parking

Research

Research 1	Research 2	Research 3
Arduino-MATLAB based Semi-Autonomous Car *Er. Vikrant Tomar, Shweta Sengar*	Using Visual Lane Detection to Control Steering in a Self-Driving Vehicle *Kevin McFall*	Design and Implementation of Autonomous Car using Raspberry Pi *Gurjashan Singh Pannu, Mohammad Dawud Ansari, Pritha Gupta*

Materials

Materials (detailed list)	Quantity (be specific)
Arduino Mega	1
Ultrasonic sensor	4
4 Wheel Motor car	1
Motor driver	1
GPS	1
Push button	1
Wires	50

Procedure

Step 1

Introduction of the topic.

Step 2

Literature Review.

Step 3

Analysis, Implementation and Results

Step 4

Conclusion

Data / Observations

- see — a hazard
- think — decide to stop
- do — to place your foot on the brake pedal until you stop.

Results

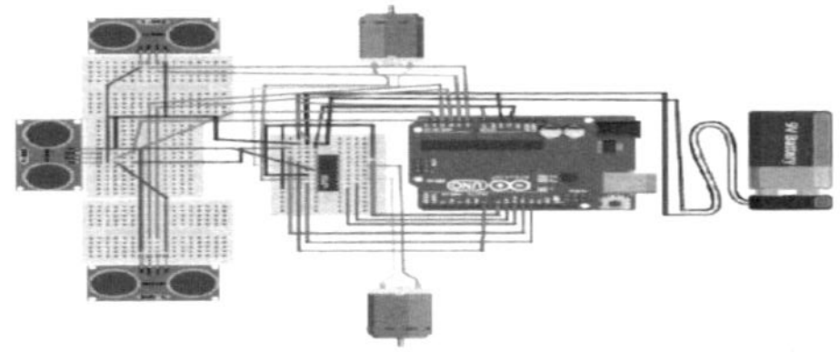

Conclusion

- First of all, self-driving cars are the future; there is no denying this point. I'm still working on my self-driving car to make it stand out among others.
- Secondly, Self-Parking car gives a constrained space, many car drivers today face a problem in parking their cars. Parking a car in a small space requires a series of forward, reverse motions and turns which prove to be a complicated task for most car drivers to handle. It will very helpful for those drivers.

Works Cited

- https://www.ee.iitb.ac.in/uma/~wel/wel45/public_html/edl09a/dd03.pdf
- http://forune.com/2015/09/22/aaa-self-parking-cars/
- http://www.businessinsider.com/autonomous-car-limitations-2016-8
- https://www.businessinsider.com.au/the-18-driverless-car-breakthroughs-2030-2016-12

Inquiry Based Learning in Higher Education and its Role in Promoting Higher Order Thinking

Amal Salim Al-Abri

Ibri College of Technology, Ministry of Manpower

Higher order cognitive skills are a necessity for the 21st century learners as they help creating critical thinkers and innovators who are able to make personal, social and economic decisions and communicate their ideas effectively. An effective way that helps promoting higher order thinking skills is active learning which is a key aspect in enhancing higher order cognitive skills among the students. Active learning is essentially defined as "students doing things and thinking about what they are doing" (Bonwell and Eison, 1991) or as Michael Prince explains, "Active learning is generally defined as any instructional method that engages students in the learning process. In short, active learning requires students to do meaningful learning activities and think about what they are doing" (Prince, 2004). For the purpose of this study, inquiry-based learning is investigated as an effective pedagogical approach for higher education learners. Inquiry-based learning (enquiry-based learning) is a form of active learning where rather than presenting facts or knowledge, it starts by constituting questions, problems or scenarios. While the students have to investigate, search for answers, evaluate information and present their findings the teacher has to facilitate the whole process. Inquiry-based learning includes problem-based learning and is usually used in small scale investigations and projects, as well as research.The inquiry-based instruction is closely related to the evolution and practice of the cognitive skills.

Keywords: 21st century cognitive skills, Problem-solving skills, Critical thinking skills, Learning through inquiry, IBL, PBL, Active Learning, Collaborative Learning, Higher order thinking Skills

B2B E – Procurement Support System

Juhaina Juma Abdulah Al-Ghailani, Sheikha Hashem Issa Alsulti, Shimaz Khan Shaik

Abstract: The procurement process in any organization can significantly influence the success of a company and currently, this process operates in a dynamic, global, complex and challenging environment. To operate efficiently and effectively it has to create the appropriate structures and make use of the ICT. In line with the above requirements, B2B based e-commerce has been growing rapidly and enterprises around the world are fast adapting this mode of commerce to enhance their business opportunities. However, the adaptation of the B2B, by the SMEs in Oman warrants a detailed study. This paper attempts to identify: the status of the B2B, and the factors that impede the growth of B2B among SMEs in Oman. Also, it furnishes the facts from the literature survey don on this topic. A model B2B system for the SMEs in Oman is proposed and developed to support the business activities such as procurement and selling.

Keywords: B2B E-Procurement system, SMEs in Oman, B2B E-Commerce, ICT

References:

1. Export.gov. (2018). Oman - eCommerce | export.gov. [online] Available at: https://www.export.gov/article?id=Oman-eCommerce [Accessed 20 Jan. 2018].

2. Ncsi.gov.om. (2018). SME status. [online] Available at: https://www.ncsi.gov.om/Pages/NCSI.aspx [Accessed 30 Jan. 2018].

3. Riyada, (2018). SME Data. [online] Available at: https://riyada.om/en-us/smeprograms/Pages/BigOpen-Data.aspx [Accessed 30 Feb. 2018].

4. Internetworldstats.com. (2017). World Internet Users Statistics and 2017 World PopulationStats.[online]Availableat:http://www.internetworldstats.com/stats.htm [Accessed 5 Dec. 2017].

5. Kbmanage.com. (2017). e-Procurement - What is it? Definition, Examples and More. [online] Available at:https://www.kbmanage.com/concept/e-procurement [Accessed 5 Dec. 2017].

6. Slideshare.net. (2017). Webinar - Oracle iProcurement in Oracle Purchasing Application. [online]Availableat:https://www.slideshare.net/iwareoracle/webinar-oracle-iprocurement-in-oracle-purchasing-application [Accessed 5 Dec. 2017].

7. Businessgateways.com. (2017). Joint Supplier Registration System(JSRS) for the Oil & GasIndustry-OmanB2BRegistrationPortal.[online]Availableat: https://www.businessgateways.com [Accessed 5 Dec. 2017].

8. Tejari, (2017). Oman Air To Reach New Heights With Tejari. [online] Omanair.com. Availableat:http://www.omanair.com/en/about-us/press-releases/oman-air-reach-new-heights-tejari [Accessed 5 Dec. 2017].

1 Introduction

E -Commerce is the process of buying, selling goods and services over an electronic network and it can significantly influence a company's success. However, this latest trend in the commerce has to be adapted by the business community to promote their business opportunities. The E-Commerce can be categorized as B2B (Business to Business), B2C (Business to Consumer), and C2C (Customer to Customer). Many business entities are embracing the B2B model of E-Commerce to benefit their business activities such as procurement and selling. There are many factors, indicate that the B2B would immensely benefit the SMEs in their business pursuit. However, the utilization of the B2B E-Commerce by the SMEs in Oman needs a comprehensive study. Hence this attempt is to find the status of the B2B utilization by the SMEs in Oman and to find the factors that impede its growth among the SMEs. According to the literature survey it was found that there are limited E-Commerce activities by the private sector in Oman (Export.gov, 2018). However, the Government of Oman is actively promoting a "digital society" and "e-government" services through the information technology authority (ITA) (Export.gov, 2018). A study by Frost &

Sullivan internet spending in the middle east is expected to reach $415 billion and the Internet users constitute 74.2 % of the total population. Oman enacted a law on electronic transaction on May 2008 to legalize the electronic transactions. In 2016, the Central Bank of Oman launched a national debit card payment gateway infrastructure for e-commerce transactions (Export.gov, 2018). The NCSI data reveals that so far 32,441 SMEs have been registered with the Government so far (Ncsi.gov.om, 2018). According to the data from the Riyada, the GDP from the SME is set to double by 2020 in Oman (Riyada, 2018). Oman companies and individuals buy products from different companies around the world and they do the payment using the bank transfers and money exchangers. Study finds that the companies in Oman use the B2B services provided by the E-Commerce sites such as: www.alibaba.com, www.aliexpress.com and these sites are not based in Oman. Also, big enterprises such as Oman air, avail the procurement consultancy services provided by the E-Commerce consultant, Tejari (Tejari ,2017) and this involves consultancy fees. Also, the study finds that the B2B services provided by some E-Commerce sites such as tradeboss.com does not provide required B2B functionalities such as RFQ, Payment gateway, Purchase order etc. This study finds that a dedicated B2B portal for the SMEs in Oman is needed to help in their procurement and selling activities. A B2B application model that encompasses all the functionalities of a B2B using FOSS based development tools is proposed, designed and developed.

Status of the B2B in Oman for the SMEs

- Currently there are no B2B system which is exclusively dedicated for the SMEs in Oman.
- Large enterprises utilize B2B procurement consultancies provided by the consultants and it involves huge cost and the SMEs are not able to afford the cost involved in the consultancy.
- The free B2B service provided by some E-Commerce sites don't provide required services of full-fledged B2B.

B2B Software Architecture

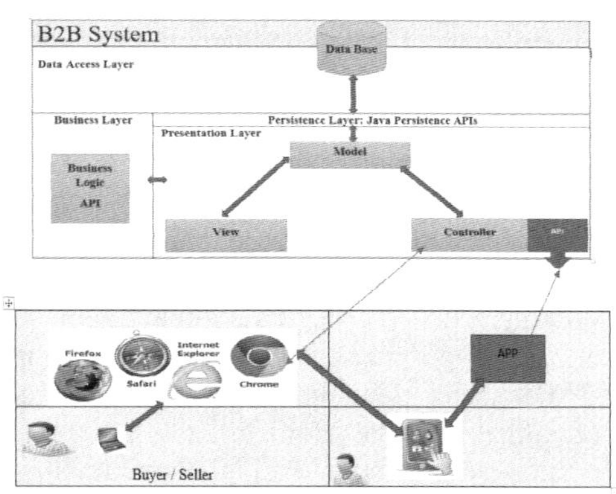

Use Case Diagram

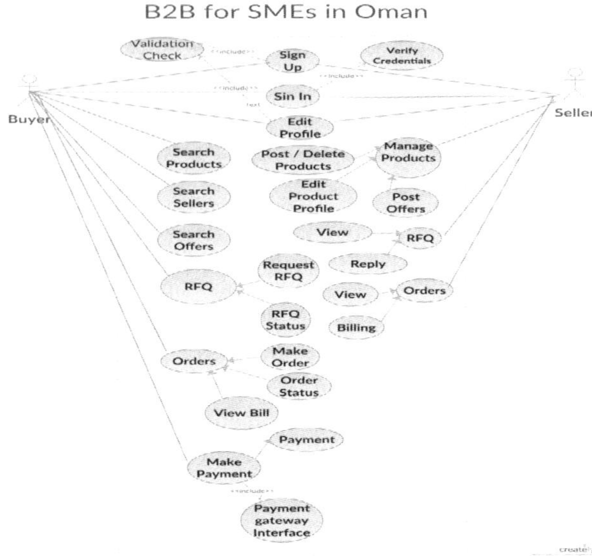

Software Used

- Java,
- Spring Framework,
- Eclipse,
- JPA,
- Thymeleaf,
- Tomcat,
- MySQL, JUNIT, HTML5, Jquery.

Sample Screen

Conclusion

Study to find the: status of the B2B system and the factors that limit the growth of B2B among SMEs was performed. It shows that a dedicated B2B for SMEs in Oman is required to facilitate the trade activities of the SMEs. A B2B system with the B2B functionalities is designed and developed that would facilitate the SMEs to perform the procurement and selling process. However, a comprehensive study is required find the requirements of the SMEs that would enhance the B2B system with more services that fulfills the requirements of the stakeholders.

Personal Health Care System Using IOT

Hatim Abdullah Mohammed Al-Hatmi, Hasna Mohammed Ahmed Al-Salti | Faisal Aga | Sur College of Applied Sciences, Sultanate of Oman

Problem Statement

- To identify chronic diseases patient suffers from sudden changes of his/her body vital states especially when the patient is alone and unattended.
- To design and implement a system which monitors and maintain the patient body vital states.

Project objectives

- To Develop a an application which addresses the challenges faced by the patient when he/she is left unattended during emergency.
- To implement PHC system in hospitals in our country, Sultanate of Oman.
- To enable the patient/doctor/friends and relatives to use this application in a user friendly manner.
- To assist the patient rapidly when he/she faces any sudden changes in their health conditions.

Project Overview

- Personal Health Care System (PHC) is an application used by any individual in emergency conditions.
- PHC provides a faster, efficient, flexible health monitoring application via web and mobile application using IOT concepts.
- PHC system provides the patients to use a bracelet which communicates faster with the mobile device and sent alerts to the already registered family and relatives.

Status of PHC System in Oman

- Health issues are raising in Oman
- People expecting best health service in Oman
- Currently there is no system which will monitor patient's vital states and send instant alerts to the doctor or family & friends in order to attend the patient in emergency situations in Oman.
- Patients health record is not accessed in time to time basis.

Closed Loop Approach Methodology

Activity Diagram

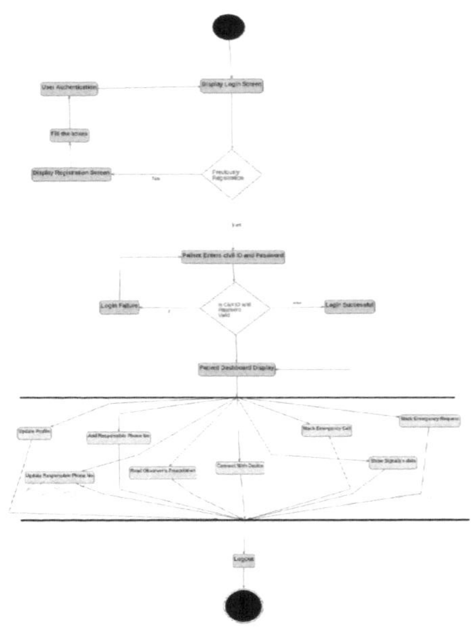

Use Case Scenarios

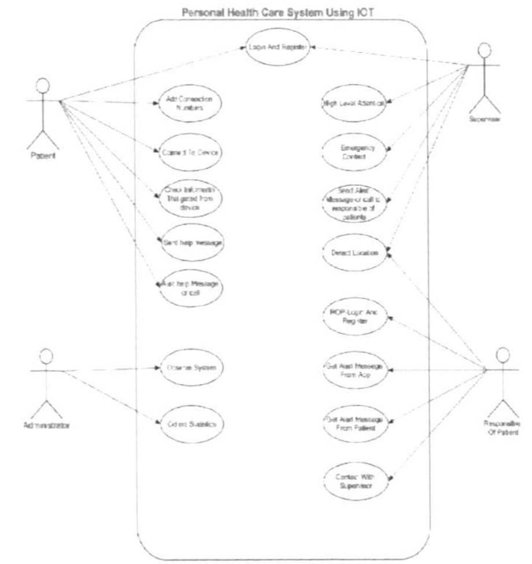

Conclusion

- Providing an intelligent solution for the patient's managing PHC in Oman.
- PHC is helpful to the doctors, nurses for monitoring patients health conditions in the form of report which includes tables or charts.

Works Cited

- "An IoT Framework for Healthcare Monitoring Systems", : https://www.researchgate.net/publication/303923009, 2016
- *Internet of Things in healthcare: Interoperatibility and security issues*. Available from: https://www.researchgate.net/publication/261116016_Internet_of_Things_in_healthcare_Interoperatibility_and_security_issues [accessed Dec 31 2017].

كـلـية الشـرق الأوسـط
Middle East College

By Rasha Alamri
11F8267

Design and Implementation of educational tools for Eco Kids Nursery, under ministry of social development

1 Aim:

To enhance the learning experience of nursery students with teaching aid, that includes fun based application.

2 Objectives:

1. Develop an interface which can be easily navigated by small children with parental control and filtering.
2. Design and implement an innovative and new application that will encapsulate all the educational requirements of a nursery in addition to providing them with an entertainment system.
3. Complete the task within a timely fashion, meeting all the deadlines, ensuring that the client is satisfied.

3 Rationale:

Main purpose is to aid in the educational development of children aged 3 months to 4 years old who attend nursery's.

4 Methodology:

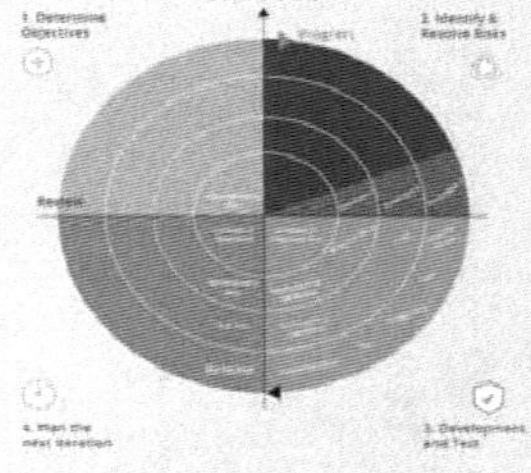

5 Clients:

Eco Kids Nursery, is under supervised by the Ministry of social development. Furthermore, It Is the first environment friendly nursery in GCC. it is located in Muscat - Al Koudh, Opened on 7/2/2017, The owner of the nursery is Khalaf bin Nasser Al-Farsi, the current Age group is from 3 months to 3 and a half years old.

6 Risks:

***External:**
- Compliance Risk and Natural Risk.

***Internal:**
- Technical Risk, Management Risk, Resource Risk and Timing Risk.

7 Limitation:

Implementing in apple store and Make it in Arabic language.

8 Time Plan:

Week	Task
Week 1	Get the idea of project
Week 2	Initiation report
Week 3	Determine project requirements
Week 4	Data Collection Method and analysis
Week 5	Submit Draft of Chapter 1
Week 6	Write literature review and feasibility analysis
Week 7	Write Data collection method and Research paper
Week 8	SWOT analysis
Week 9	Submit Draft of Chapter 2
Week 10	Resource Plan, Data analysis and Initial Block Diagram
Week 11	Project requirements
Week 12	Choosing the methodology
Week 13	Submit Draft of chapter 3
Week 14	Submit Final report of planning
Week 15	Final presentation of planning
Week 16	Start discuss the ideas and meet the clients
Week 17	Discussion about system design
Week 18	Start Designing
Week 19	Discus interface design
Week 20	Write Chapter 4
Week 21	Submit the draft of Chapter 4
Week 22	Feedback from the supervisor
Week 23	Coding
Week 24	Project testing
Week 25	Submit the draft of Chapter 5
Week 26	Feedback from the client
Week 27	Submit the draft of Chapter 6
Week 28	Processing poster, report, Presentation and complete project
Week 29	Submit the final project
Week 30	Presentation

9 Refrences:

1. Journals.sagepub.com. (2017). Putting Education in "Educational" AppsPsychological Science in the Public Interest - Kathy Hirsh-Pasek, Jennifer M. Zosh, Roberta Michnick Golinkoff, James H. Gray, Michael B. Robb, Jordy Kaufman, 2015. [online] Available at: http://journals.sagepub.com/doi/abs/10.1177/1529100615569721 [Accessed 2 May 2017].
2. Rauterkus, M., Peter W A Wood, P. (2017). 1001 things to do with LiveCode. [online] Livecode1001.blogspot.com. Available at: http://livecode1001.blogspot.com/ [Accessed 11 May 2017].
3. Holgate, C. and Gerdeen, J. (2015). LiveCode mobile development beginner's guide. 1st ed. Birmingham, UK: Packt Publishing.
4. Lavieri, E. (2013). LiveCode Mobile Development Hotshot: Create Your Own Exciting Applications with 10 Fantastic Projects. 1st ed. Packt Publishing.
5. Nngroup.com. (2017). 10 Heuristics for User Interface Design: Article by Jakob Nielsen. [online] Available at: https://www.nngroup.com/articles/ten-usability-heuristics/ [Accessed 20 May 2017].
6. Ecokids.ca. (2017). Play and learn! □ Eco Kids. [online] Available at: https://ecokids.ca/blog/take-action/play-and-learn [Accessed 20 May 2017].

Understanding and Mapping 21st Century Competencies in the Framework of Higher Education, Employability and Life

Dr. Madhuri Goswami
Associate Professor, Department of English, Government Postgraduate College, Chomu, (303702), Jaipur, Rajasthan, India

Abstract

The paper has sought to investigate 21ST century skills and map the significance of these skills in the pursuit of professional awareness by educators as well as learners. The present digital age demands for the professionals embellished with soft skills, interpersonal skills, communication skills, critical and creative thinking, and problem solving skills in place of the masters with merely discipline-specific knowledge. The study has identified the factors and their dimension which affect skills in the learning process. It has also dealt with the purview of communication skills in the perspective of education and employability. One class of an undergraduate course in a suburban town in Rajasthan, India was selected as the research site and the sample 20 student participants were provided the training cum learning workshops to assess and sharpen their employability skills. The workshops offered various task based activities- to be accomplished collaboratively as well as individually. Some project based learning activities, talks, and mock interviews were organized and learning tools and Apps were provided to the target group by the Entrepreneurship and Employment Cell in the collaboration of government and Commissionerate of Higher Education. Most of the participants were actively engaged in the training sessions, but the pressure to complete the regular curriculum timely and constraint of time kept many of them at the passive mode. The research finds that the educational and professional programmes, helping learners acquire life skills, are the need of the hour. Being graduated or obtaining a degree and discipline specific knowledge do not fetch and condition personal, emotional and social skills. The great urgency of integration of skill-focused activities into the university or college curriculum and incorporation of technologies and PBL activities into teaching on the part of educator are significant outcomes of the study. Developing easily accessible tools, Apps, podcast, etc. to ameliorate learners' soft skills is another substantial step in the process. Effective strategy to inculcate learners with life skills, positive attitude and confidence is an essential provision for transcending the professionals from local to global strata and preparing them to cope up with the 21st century challenges.

Keywords: **Communication, collaboration, life skills, higher education system, pedagogy**

Introduction

The present scenario of life, work and citizenship has evolved a demand for a set of competencies comprising skills, abilities and learning dispositions requisite for skilled populace. Educators, corporate, academics and governmental agencies have acknowledged these skills as significant and indispensable for success in 21st century society and workplaces. Basic skill competencies and subject-specific knowledge are not suffice to make learners competent enough to communicate, share and use information to solve complex problems of the present information age. The transforming backdrop of education has been exponentially engaging the discussions around the world. The responsibility of education systems has been defined as preparing students for an immediate context as active and productive citizens (Dellamore, 2004). The learners need to be innovative, creative, enterprising and adaptable to be able to respond to the 21st century's social, economic, civic, and health needs. Ledward and Hirata (2011) reflect on the shift over the last century, "Shared decision-making, information sharing, collaboration, innovation, and speed are essential in today's enterprises" (quoted in Pacific, 1).

Defining 21st century skills

The concept of *21st century skills* has been widely open to discussion and interpretation around the world. There are piles of cross-disciplinary, relevant and current definitions of *21st century skills*. In general, the term '21st century skills' refers to a cluster of specific competencies, work habits, inclinations, character traits, content, knowledge and literacy considered critically important to success in the rapidly changing digital society, in the context of higher education, contemporary careers, workplaces and life. Effective communication, collaboration, innovation, connection, cognition, and creative and critical thinking, are reckoned, with consensus, as the foremost components of the set of 21st century skills. Tony Wagner, in the book *The Global Achievement Gap* (2008), proposes the following seven survival skills:

1. Critical thinking and problem solving
2. Collaboration and leadership
3. Agility and adaptability
4. Initiative and entrepreneurialism
5. Effective oral and written communication
6. Accessing and analyzing information
7. Curiosity and imagination.

These competencies have been, more or less, acknowledged as necessary for success in the 21st century education, career and life. "…a crux of success or failure as a society is to know which core values to hold on to, and which ones to discard and replace with new values, when times change" (Diamond, 2011, pp. 433). Globalization and massive global migration have transformed local markets and workplaces into the global ones bringing in the demand for professionals with the abilities to address new challenges and to adapt to new environments. Thus, today's globally interconnected cultures and economies have set the necessity for students to learn how to collaborate, communicate, analyze, and solve problems with people beyond national boundaries (Appiah, 2006). The Organization for Economic Cooperation and Development (OECD) *Report: Preparing Teachers and Developing School Leaders for the 21st Century* states that, "Nations around the world are undertaking wide-ranging reforms to better prepare children for the higher educational demands of life and work in the 21st century" (Schleicher, 2012, pp. 3). Of course, there is widespread consensus that the responsibility to adequately prepare our students with these survival skills is due with the education systems and the higher education system is considered a key agent in educating and assessing the current generation to cope up with new challenges. Griffin and Care (2012) also support the view, "New conceptions of educational standards and assessment…are a key strategy for accomplishing the necessary transformation" (p.18)

Role of Educators

Teaching 21st century skills is not just teaching but enabling students to apply these traits or dispositions in real-life situations and having competencies to teach essential skills. Teacher-educators have a significant role of the facilitators to guide student-learners to become gradually self-guided and motivated. The real challenge for the educators is to model or carve a desire to "learn, unlearn and relearn" among learners to prepare them for local as well as global interaction (Toffler, 1991). Leadbeater (2008), a leading innovation thinker, states that the successful reinvention of educational systems worldwide can take place through transforming pedagogy and redesigning learning tasks. He, further, suggests promoting learner autonomy and creativity as part of the solution. So, educators, in particular context of higher education, need to prepare the graduates to possess, learn and assess 21st century knowledge and skills, necessary not only for career success but for personal and civic quality of life as well.

The study has observed that following factors exercise a deep influence on the teaching and learning process of the acquisition of skills:

1. Identification of learning styles and teaching accordingly
2. Innovative approaches and methods used by educators
3. Making knowledge interesting, productive and relevant
4. Instilling the notion of self-reliance among learners

Method of research

The present study has worked upon a project to develop and sharpen the essential skills of college students in the context of employability and life. To explore the optimal model for developing 21st century skills, the Innovation and Entrepreneurship Cell of the Government Postgraduate College, Chomu, located in a sub-urban town of Rajasthan, India, designed a co-curricular programme to work upon the skills of the students. The Cell, randomly, selected twenty participant students from the graduation courses of Arts, Commerce and Science streams and conducted the study with the following guiding principles:

1. Provide learners the opportunities for developing soft skills
2. Enhance the employability skills of learners
3. Engage learners in discovery process and to make them able to adapt and innovate in response to new demands and shifting circumstances
4. Inculcate Information and Communication Technology skills in learners.

The study adopted three pedagogical principles – *personalization, participation and productivity* stated by McLoughlin and Lee (2008a). It sought to locate the personal requirements of the selected learners in the first stage; in the next stage, it ensured the active participation of the learners in the skills-development programme through GDs, training workshops, Apps, tools and talks; and the final stage examines the productivity of the target group.

In the initial stage of the project, the learning needs and different learning styles of the learners were identified to design the activities for them. Though, it seemed quite difficult to make out which particular learning methods and pedagogy would exercise constructive impact on an individual learner. However, the Cell opted for pedagogies that supported deeper learning inclusive of collaborative and project-based learning. The target group was evaluated through group-discussions and a rubric in order to find out what aspects of skills needed to be worked upon to increase entrepreneurial and employability skills. The 21st Century Student Self-Assessment Rubric, developed by TORCH programme, USA was used to make learners identify and work upon the relevant set of skills. This tool proved quite helpful for both educators and learners; the educators could discover the individual strengths and needs of learners; the learner participants were brainstormed and enabled to comprehend the dimensions of 21st century skills. After discovering learners' individual strengths and needs, communication and analytical skills, critical and creative thinking, collaboration, and technology literacy were found as key areas to work upon in this project.

Saavedra and Opfer (2012) have aptly suggested that 21st century pedagogy must employ innovative and research supported teaching strategies, learning technologies and real world applications. A series of skill-based training workshops were organized with intent to prepare students for thinking critically and creatively and to enable them to communicate effectively, to collaborate with others, and to develop their employability skills. Productive and rationale thinking is the core of effective learning. The learners acquire a deep understanding of the methods which they can use to fix problems and deal with unfamiliar situations through practical application of the knowledge of thinking skills. The increasing practice of thinking strategies is quite helpful to bring learners' autonomy, in addition to self-motivation and confidence. The content used in these workshops served as a means of teaching communication skills and critical thinking. The project sought the service of Indian Institute of Technology, Bombay, India, Hello English Premium and *Dishari* (e-learning programmes launched by the government of Rajasthan), a rubric tool, subject-experts, motivational speakers, and faculty members to create and develop a range of engaging pedagogies that increases student participation, collaboration, and independent thinking.

The participant learners were supplied the exposure to various significant issues such as Personality Development, Time and Stress Management, Entrepreneurship and Life Skills, Career Counseling, Job Application and CV Writing and mock interviews to identify, comprehend, render, create and apply through training workshops comprising interactive presentation by subject-experts and reflective feedback by participants. Communication skills, be it verbal, non-verbal and written communication, contribute to produce harmonious relations at workplaces among co-workers and between employees and customers. In order to enhance communication skills, the Cell rendered the service of a project, Upskill Proficiency in English for Rajasthan (UPER), launched by the Government of Rajasthan to provide an English learning App- Hello English Premium to the students. The target group was provided the App to make them fluent in English, thereby improving their verbal and written communication skills. All four skills- listening, speaking, reading and writing- were taught and practiced through the interactive modules, activities, news articles, stories and dictionary which were part of the App.

Collaboration, both real and digital, is one of the significant assets and essential to success in any corporate or business milieu. Team work hones the ability to think critically encouraging cooperation, adaptation, and tolerance for divergent opinions and ideas. The traditional transmission method of teaching minimally allows students to learn from each other and develop the teamwork skills. Therefore, to teach the collaboration, the target group was assigned a task of preparing a project-file by each sub-group of four learners describing their learning experiences and outcomes. This gave the learners a prospect to share their thoughts, opinions, and ideas productively.

The use of information technology has been infiltrating the peripheries of higher education, employment and life. The learners as well as educators need to be tech-savvy to successfully negotiate the demands of the 21st century. Apart from having pedagogical value, technology offers a platform "to develop their problem solving, critical thinking, and communication skills; transfer them to different contexts; reflect on their thinking and that of their peers; practice addressing their misunderstandings; and collaborate with peers—all on topics relevant to their lives and using engaging tools" (Saavedra and Opfer, 2012, p. 16). In order to make students enable to learn digital literacy and enhance IT skills, Spoken Tutorials were provided to the learners with the help of Indian Institute of Technology, Bombay. The Cell got these participant learners registered with IIT, Bombay to access and learn the open source technology software programmes such as, Web Designing, and Libre Office Suite etc. The e-learning Apps such as, Hello English Premium and *Dishari,* offered various activities to improve general awareness, reasoning, mental ability of learners to enhance their employability skills, critical thinking, and analytical skills.

The workshops, task based activities, inspirational talks, and meeting with successful alumni were organized during the project with an aim to expand the classroom experiences of the students and learning Apps and tools were provided to learners to harness the learning opportunities offered by digital technologies. The final stage of the project comprised an assessment. The assessment criteria were designed to make learners capable for knowledge application. The aim of the assessment had been to map the dimensions of the skills from the aspect of execution and knowledge application and to ensure that the skills which they had learnt be applied to the real life situations.

Assessment Criteria

1. Group Project File: A group of four students prepared a file containing three parts- a) The introduction of project, b) Relevance of hand-outs, and c) Reflection. This task was assigned to develop and assess collaboration, creative thinking and analytical skills.

2. Oral presentation: Each student from the group was given an opportunity to present their learning experiences and display their capability to use presentation aids. This task was assigned to develop independent learning and to reflect on their learning and take appropriate action to improve it.

3. Written Report: Each learner submitted a written report of their learning outcomes. It aimed to enable them to learn independently through self-reflection and evaluation of their own work processes

RESULTS

The study found that the project of Entrepreneurship and Employment Cell helped learners to a significant extent to identify and harness employability skills, namely, collaboration, digital literacy, communication skills, and analytical skills. The participant students learned and practiced the critical reflection and analytical skills as they posed questions, researched, analyzed, evaluated and communicated information, concepts and ideas during task-based learning activities. The use of apps and tools helped learners to practice active listening, use appropriate language and vocabulary, and give effective oral presentation. The project with its various implements, such as, motivational talks and task based learning, inculcated exemplary zeal, civic sensibility, collaborative spirit, and non-verbal communication skills among the learners that was noticed in the oral presentations and group project file. It also enabled them to practice various competencies in real life situations. The project, in its final stage, visualized the target group equipped with the ability to apply class-room gathered knowledge to real-life contexts. The active learners showed heightened motivation in class, improved reading and writing skills, and enhanced engagement in the learning. Twelve out of twenty learners got placement at the placement and job fair organized by the institution with the support of various corporate agencies.

However, the study confronted few challenges as the pressure to complete the main university curriculum timely and constraint of time kept many of the participants at a passive mode in the activities of the project. The study suggests that the inclusion of skill-development courses in the regular curriculum and inter-disciplinary study in place of stand-alone courses would bring a substantial solution to it. It also suggests that the students or learners should be given a comprehensive understanding of 21st century skills through brainstorming before the initiation of teaching essential skills. This comprehension turns to be the spur of self-guidance and motivation and fosters creativity of learners and they can successfully perform in educational, professional and other life contexts through such creativity, collective participation and implementation of learnt attributes in real-life situations.

The study compared output of the group of participant learners with that of other students and noticed that being graduated or obtaining a degree and discipline specific knowledge do not ensure the procurement of personal, emotional, social and life skills. It finds the great urgency of integration of skill-focused activities into the university or college curriculum and incorporation of technologies and PBL activities into teaching on the part of educator. In order to implement these findings, there is a need of the effective professional development of educators to teach 21st century skills, and it reckons on appropriate materials, activities and sustained practice of new techniques and information. Hence, the concept of innovation, applied to both curricula and teaching methods, has come up as quintessential to effectively teach these skills.

Conclusion

The changing pace of current century has transformed facets of teaching and learning and redefined the purpose of the higher education system and the role of educators. The widening diversity of 21st century challenges of career, workplaces and life can be addressed through using innovative teaching materials along with the conventional transmission methods of teaching and learning. It is hypothesized that the inclusion of teaching soft skills in academic curriculum, in addition to hard skills, would sharpen the learners' employability skills such as innovation, collaboration, analytical and enterprising skills, and critical and creative thinking. For this purpose, the Cell, after conducting the project, has initiated a collaborative venture with IGNOU, New Delhi to add skill-development course in the main curricula. Though, its effectiveness would be explored when the duration of the course is over. However, the constructive effect of the project strongly recommends for this innovation. To conclude, the shifting paradigm of higher education from curriculum-centered to learner-centered approach expects from teachers to readily accept diversity with new pedagogical practices and foster new forms of education systems. Effective teaching strategies to inculcate learners with life skills, positive attitude and confidence is sine quo non for transcending the professionals from local to global strata and preparing them to cope up with the 21st century challenges. The incorporation of interdisciplinary learning, the weaving of specific skills in all content areas and the integration of skill-based curriculum in the main prescribed college or university curriculum would prepare a skilled workforce and sensible citizens in addition to bringing equity in the educational opportunity.

REFERENCES

Appiah, K. A. (2006). *Cosmopolitanism: Ethics in a world of strangers*. New York, NY: W.W. Norton & Company.

Dellamore, E. (2004). "Imperative". *Pathways to Thinking Schools*. Eds. Hyerle, David N. and Alper, Larry. California: Corwin Press. pp. 26-54.

Diamond, Jared, M. (2011). *Collapse: How Societies Choose to Fail or Succeed*. New York: Penguine Books.

Griffin, P. and Care, E. Eds. (2012). "The ATC21s Method". *Assessment and Teaching of 21st Century Skills: Methods and Approach*. Melbourne, Australia: Springer. pp. 3-33.

Leadbeater, C. and Wong, A. (2010). *Learning from the Extremes: A White Paper*. San Jose, Calif., Cisco Systems Inc. www. cisco.com/web/about/citizenship/socio-economic/docs/ Learning fromExtremes_WhitePaper.pdf [Accessed 24 November 2017].

McLoughlin, C. and Lee, M.J.W. (2007). Social software and participatory learning: pedagogical choices with technology affordances in the Web 2.0 era. ICT: Providing Choices for Learners and Learning: Proceedings Ascilite Singapore 2007, pp. 664-675. /www.ascilite.org.au/conferences/ singapore07/procs/mcloughlin.pdf/ [Accessed 19 October 2017].

Saavedra, A. and Opfer, V. (2012). Teaching and Learning 21st Century Skills: Lessons from the Learning Sciences. A Global Cities Education Network Report. New York, Asia Society. Web. http://asiasociety.org/files/rand-0512report.pdf [Accessed 10 December 2017].

Schleicher, Andreas. (2012), Ed., Preparing Teachers and Developing School Leaders for the 21st Century: Lessons from around the World, Paris: OECD Publishing. https://www.oecd.org/site/eduistp2012/49850576.pdf [Accessed 17 December 2017]

Toffler, A. (1991). Powershift: Knowledge, Wealth, and Violence at the Edge of the 21st Century. New York: Bantam Books.

Triling, B. and Fadel, C. (2009). '21st Century Skills: Learning for Life in Our Times'. Jossey-Bass, San Fransisco. CA 89-91.

Wagner, Tony. (2008). *The Global Achievement Gap: Why Even Our Best Schools Don't Teach the New Survival Skills Our Children Need- and What We Can Do About It.* New York: Basic Books.

Acknowlegement: TORCH Programme, New York and CIN for providing 21st Century Student Self-Assessment Rubric. /https://21clearninginua.wikispaces.com/file/view/z8Z_dq52.pdf [Accessed 06 September 2017]

Biographical Notes: Dr. Madhuri Goswami, Ph.D., CELTA, is an Associate Professor of English in the Department of English of Government Postgraduate College, Chomu, Jaipur, Rajasthan, India. She has been teaching English Literature and Language with government colleges since 2002. Her research and publication deal with American Literature, Black writings and Contemporary Indian Writings. She has to her credit six research articles published in the international refereed research journals. She has presented papers at various international and national conferences and seminars. Her research interest includes new techniques and strategies to teach ESL and Computer Assisted Language Learning.

Deep Learning Enabled Chabot: A Tour Assistant

Sahar Mubarak Juma Al-Saadi, Buthaina Said Al-Hajri, Shameer Mohammed

Sur College of Applied Sciences, Sultanate of Oman

Abstract:

In term of tourism, online customers can perform many activities, such as booking a flight or hotel, by using normal websites or applications. Although, many of customers prefer personal services which provide a live help facility for them. This becomes available nowadays by using a Chatbot, which use deep learning technology to create a virtual conversation between a tour assistant, which is a system, and a customer. Chatbot technology is capable to deal with the customers' requests easily, quickly, and 24 hours every day. Most online users today use chatting applications more than others; this is other feature that makes a Chatbot more interesting and acceptable by customers.

Keywords:

Natural Language Processing, Chatbot, Tour Assistant, NLP, LUIS

Reference:

Bahdanau, D., Cho, K., and Bengio, Y. (2015). Neural machine translation by jointly learning to align and translate. In ICLR.

Liu, C.-W., Lowe, R., Serban, I. V., Noseworthy, M., Charlin, L., and Pineau, J. (2016). How NOT to evaluate your dialogue system: An empirical study of unsupervised evaluation metrics for dialogue response generation. arXiv:1603.08023

Serban, I. V., Lowe, R., Charlin, L., and Pineau, J. (2015). A survey of available corpora for building data-driven dialogue systems. CoRR, abs/1512.05742.

Li, J., Galley, M., Brockett, C., Gao, J., and Dolan, B. (2016). A diversity-promoting objective function for neural conversation models. In NAACL.

1 Introduction

1.1 Project overview:

Tourism has recently become one of the significant economic industries in Oman. To enhance that, Oman will develop a roadmap for tourism sector for the next 30 years as one part of its strategy for tourism. Tourism in Oman is contribute to promote economic diversification, protect the environment and maintains the nation's cultural integrity. Technology plays a significant role in the tourism field. One of its great roles that it allows tourism companies or organizations to replace expensive human labour with technological labour, this led to reduce labour costs and avoid many of customer service issues. Today people use the technology more than ever before, which makes the tourism companies and organizations take advantage of this to keep communicate with the tourism customers. One of the intelligent technology that used by tourism industry is Chatbot. Chatbots was unknown on the beginning of this century; however, they become increasingly popular today and in various sectors like tourism and medical. We can define Chatbot as a system that imitates the human conversation through text chat, voice command, or both. It uses deep learning technology, which allows it to be more flexible in term of accepting the user inputs and working well even though it has a very limited knowledge domain. Chatbots are playing a role as a personal assistant. Many companies and governments are using them to provide a personal help for their customers.

1.2 Problem Statement

- Difficult for the tourism customers achieve particular need directly from search engines, where they got many results to check.
- Getting particular need take long time by using normal tourism websites or application.
- Search engines are without feeling, which make the customer bored while searching.

1.3 Purpose

- Replace expensive human labour with technological labour.
- Create system available anytime for the customers and can handle multiple costumers at the same time.
- System should give the response for the given request directly and quickly.
- System must have great knowledge about tourism.
- System should use artificial intelligence technology to minimize the bot misunderstanding.
- System should be friendly, easy to use and mimic the human conversation while chatting with customers.

1.4 Project Objectives

The objective of the project is to create a system that enhance the role of tout assistant by provide a personal help for tourism customers and provide quick, reliable and direct answer for their questions.

1.5 Scope

Chatbot contains the sensitive information about tourism in Oman. This information is helpful for people who want to know about Oman and for tourist who want to visit Oman.

The Chatbot works as any conversation applications but the deference is that one side of the conversation is a bot. The bot will receive the request from the customer, and then it will try to understand it correctly and finally provide the response to the customer. This will done by converting the customer request sentence to a query to be understood by the bot, the bot will connect with the dataset which is a database contain all the possible questions, that could be asked by customers, and their answers. Chatbot going to find the appropriate answer query for the customer request and finally return the response to the customer in a form of human understandable sentence.

1.6 General constrains of the project

- People in general have less knowledge about Chatbot technology.
- Our Chatbot support English language only, which may not be accepted by many Arab tourists.
- Technically, Chatbot is not able to decipher if the tourist try to contain or mix different languages in a single request.

1.7 Challenges

The Chatbot must be comprehensive to meet all the customer needs, which means we have to spend time and labour to collect as many as possible information about the tourism in Oman which is not easy for us ,as two students, as long as we have short time to do this. In addition, the information is updated continuously. We cannot get the new update until we ask for them from the ministry of Oman in Muscat, which will take time to provide them while we have to exploit the time as well.

1.8 Proposed System

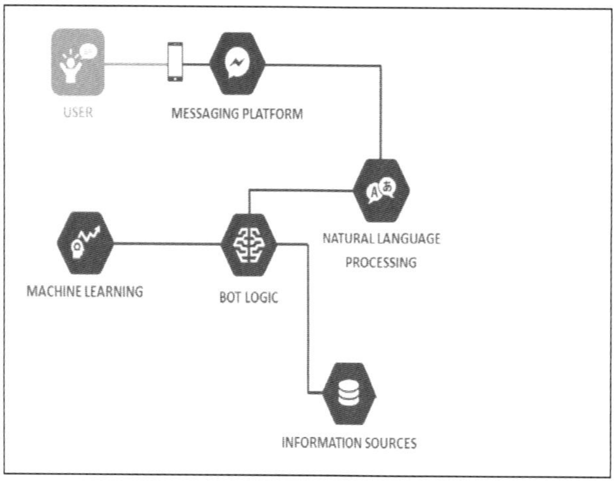

1.9 User Interface

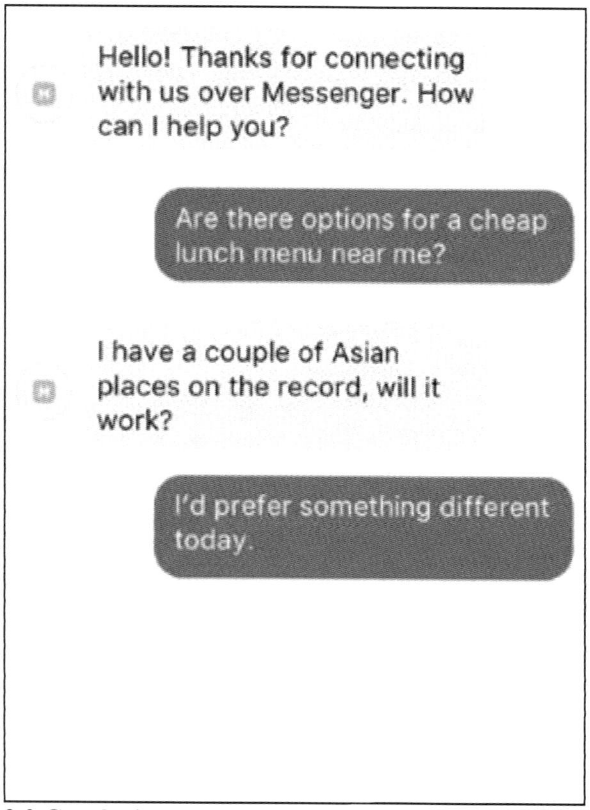

2.0 Conclusion

- Using Chatbot in tourism area will be helpful for increasing the economy of Oman.
- Implementing Deep learning techniques enhance the answering capability of the Chatbot.
- Virtual conversation between tourist and the system help for planning a time without wasting too much time on searching.

Journal of Small Business & Entrepreneurship

Knowledge Management and Organisational Performance: The Moderating role of Technology Orientation

--Manuscript Draft--

Full Title:	Knowledge Management and Organisational Performance: The Moderating role of Technology Orientation
Manuscript Number:	
Article Type:	Special Issue Article
Keywords:	Knowledge Management; Technology Orientation; Organisational Performance
Manuscript Classifications:	C. Strategic management and change in an SME; E. Special topics in entrepreneurship and SME management; Entrepreneurship and sustainable development; SMEs and regional, urban, rural, and national development; Strategic Management in SMEs
Abstract:	This study looks at technology orientation (TO) and its moderating effect on the relationship between knowledge management (KM) practices and performance of manufacturing SMEs with a focus on whether or not performance of SMEs positively relates to KM and TO, and also whether or not TO moderates the former relationship. A quantitative empirical approach was used on segregated / stratified manufacturing SMEs in Oman on the basis of "number of SMEs" and "proportionate stratified random sampling". With the notion that KM and technology have an impact on organizational performances, it was concluded that there are needs: for support and capability development with regards to R&D within SME sector; collaboration encouragement among SMEs; Government legislation review and support towards the SMEs; well-focused capacity building plan; and strive for market competitiveness. Also, ensuring appropriate adoption of emerging technologies, better understanding of the business environments, saving time-to-markets, promoting learning for enhanced problem-solving skills, and encouraging integration of new knowledge to stimulate creative products and services, may mitigate the impending problems thus could reduce costs and strengthen overall performances. Furthermore, policies need to emphasize on knowledge and technology resources within the manufacturing strategy processes and performance.

Knowledge Management and Organisational Performance: The Moderating role of Technology Orientation

Dr. Muhammad Saqib*, Robin Zarine, Shyamala Srinivas
Computing Department,
Middle East College, PC 124-Al Rusayl, Muscat, Sultanate of Oman
msaqib, robin, shyamala @mec.edu.om

About Authors

Dr. Muhammad Saqib is a Senior Lecturer and Program Manager at Computing Department of Middle East College with research interest in Knowledge Management, Innovations and technology orientation.

Robin Zarine is a Senior Lecturer at Middle East College with research interest in knowledge management, technology and education.

Ms. Shyamala Srinivas is a Senior Lecturer and Head of CAP in Middle East College, having more than 18 years of teaching experience. Her research area is Knowledge Capturing, Application, Sharing and Creation.

Knowledge Management and Organisational Performance: The Moderating role of Technology Orientation

Abstract

SMEs are believed to be the economic pillars for growth, development and socio-economic thrust. They are exemplary performers of innovative practices and great creators of employment and wealth. Unfortunately globalization and the knowledge-intensive economy are challenging their existence through a number of factors thus the recognition for knowledge acquisition and management to achieve and sustain a competitive advantage.

This study looks at technology orientation (TO) and its moderating effect on the relationship between knowledge management (KM) practices and performance of manufacturing SMEs with a focus on whether or not performance of SMEs positively relates to KM and TO, and also whether or not TO moderates the former relationship. A quantitative empirical approach was used on segregated / stratified manufacturing SMEs in Oman on the basis of "number of SMEs" and "proportionate stratified random sampling".

With the notion that KM and technology have an impact on organizational performances, it was concluded that there are needs: for support and capability development with regards to R&D within SME sector; collaboration encouragement among SMEs; Government legislation review and support towards the SMEs; well-focused capacity building plan; and strive for market competitiveness. Also, ensuring appropriate adoption of emerging technologies, better understanding of the business environments, saving time-to-markets, promoting learning for enhanced problem-solving skills, and encouraging integration of new knowledge to stimulate creative products and services, may mitigate the impending problems thus could reduce costs and strengthen overall performances. Furthermore, policies need to emphasize on knowledge and technology resources within the manufacturing strategy processes and performance.

Keywords— Knowledge Management; Technology Orientation; SME; Organizational Performance

I. INTRODUCTION

SMEs are thought to be the most dynamic firms when it comes to catalyzing economies. Organisation for Economic Cooperation and Development defines SMEs as small and medium-sized enterprises (SMEs) that are non-subsidiary, independent, and employ less than a given number of employees which varies across countries (OECD Policy, 2000). They are believed to be the economic pillars for growth and development leading to positive socio-economic performance. They are exemplary in terms of performance through innovative practices and use of new technologies (Shackelford, Adam, and Peter Yih-Tong Sun, 2009) with great capabilities to create employment and wealth when performing well (Sandberg, Karl W., 2003).

Globalization has increased the market complexity and uncertainty, has intensified the need to be more responsive and has encouraged employee mobility. These are proving to be challenges for SMEs and they seem to keep mounting through economic issues such as difficulties in securing initial capital, accessibility to resources and access to new technologies for innovation (Günter Verheugen, G., 2005). Globalization has also progressively intensified competition making safe-keeping of acquired knowledge and adoption of innovative practices with appropriate business strategy to improve competitiveness unavoidable.

It is the ultimate goal of every organization regardless of size to continue to perform, grow and progress so to ensure return on investments of shareholders (Pirnea, Ionela Carmen, and Elena Alexandra Caldararu, 2012) and it is no exception for SMEs. But due to continuous shift towards globalization as highlighted above, markets keep becoming more complex and instable and organizations are experiencing record high employee turnover while facing increasingly demanding customers (Becerra-Fernandez, Irma, and Rajiv Sabherwal 2014). Consequently, for many organizations including SMEs, survival is becoming more of the goal rather than growth threatening economic progress. This should be viewed as a significant economic threat and therefore appropriate solutions need to be identified and implemented.

Understanding the current economy is always a sensible starting point to better prepare organizations to be more responsive. The current economy is often referred to as the knowledge-based economy as currently economic productions and services are mainly based on knowledge-intensive activities (Powell, Walter W., and Kaisa Snellman 2004). Powell and Snellman further explained that in this knowledge-based economy, technical and scientific advancement may be accelerated which is the current scenario in SMEs but simultaneously they can also become obsolete much quicker. As a result, organizations end up with shorter resources value span thus requiring them to be even faster in decision making and more responsive in terms of offering new product and services. Keeping up with accelerated technical and scientific advancement investments that are becoming obsolete quicker is also becoming a hindrance for SMEs growth and progress.

It is argued that many determinants may influence organizational performance operating within this current economy and one that is being increasingly recognized as essential is knowledge management (Liao, Shu-hsien, and Chi-chuan Wu 2009). Knowledge management is emphasized upon as one of the key influential factors with regards to organizational performance and knowledge as strategic asset of organizations. Knowledge is considered as organizational power which plays a crucial role towards the survival and sustainability of organizations within this competitive business environment (Skyrme, David 2012). Therefore, to effectively respond to the ever-changing market needs, knowledge and the management of knowledge are becoming necessities to any organization regardless of its size even though the managing approach could be different (RIZEA et al., 2011).

II. Literature Review

KM is defined in a variety of ways. Some include making effective use of available knowledge resources within organizations by transforming individual knowledge into organizational knowledge (Rasula, Jelena, Vesna Bosilj Vuksic, and Mojca Indihar Stemberger 2012), systematic process of acquisition, creation, refinement, storage, transfer, sharing, and utilization of knowledge to improve employees' understanding (King, William R 2009), bringing employees together to promote innovative collaboration and be more productive from an individual and collective standpoint (Dalkir, Kimiz 2013), or a quick and efficient way to managing intellectual property (Hamzehi, Salimi and Khodayari 2016). It is simply doing what is needed to get the most out of knowledge resources and making important knowledge available wherever and whenever it is needed across organizations. Consequently, continue to expand employees' productive capabilities and their potential contributions leading to better organizational memory, quicker problem solving capabilities, more innovative, better knowledge embedded products and services, more competitive and stronger drive to the overall organizational strategies (Dalkir, Kimiz 2013).

SMEs strive on their know-how making their knowledge assets invaluable in terms of getting insights on the best possible strategies to achieve set targets. Their successes relate to properly managing their knowledge (Dollinger, Marc J. 1984; Dollinger, Marc J. 1985; Brush, Candida G., and Pieter A. Vanderwerf, 1992) which serves as a powerful element within the process or product development (Davenport, Thomas H., and Laurence Prusak, 1998). It has been noted that the absence of knowledge culture and marketing skills are key causes of poor performance of SMEs with many failing due to negligence and inadequate knowledge management (KM) practice (Egbu, Charles O., Subashini Hari, and Suresh H. Renukappa 2005; Sabri, Hala 2005; Daud, Salina, Rabiah Eladwiah Abdul Rahim, and Rusnita Alimun 2008). Therefore, it needs to be emphasized that KM has become an important driving force for achieving organizational aims and economic objectives. Furthermore, empirical research studies by Cerchione et al; 2015 showed that the use of KM practices can contribute to the overall growth of SMEs by simultaneously and significantly enhancing their performances.

Other key issues faced by SMEs are getting enough funding from banks and other financial institutions, the availability of long term loans with favorable interest rates, the collateral requirements and formalities, and the lack of innovations and training (Murthy, Y. Sree Rama 2009; Belwal, Shweta, Rakesh Belwal, and Fatema Al Saidi 2014; Atef, Tamer Mohamed, and Masooma Al-Balushi 2015; Bilal, Zaroug Osman, and Nawal Said Al Mqbali 2015). Excessive bureaucracy and transportation cost also play a major role

(McWilliam, Tyler Philip Eldridge 2011; Thorpe, Richard, Robin Holt, Allan Macpherson, and Luke Pittaway 2005).

The table below shows the percentage share of SMEs in employment, in GDP and no of companies.

Table 1:
Country-wise percentage share of SMEs in Employment, GDP and No. of Companies

Country Name	SMEs Share in Employment	SMEs Share in GDP	SMEs Share in Total no of Companies
USA	50 %	52 %	98 %
EU	67 %	58 %	99 %
China	75 %	60 %	99 %
Russia	51 %	17%	92 %
Kazakhstan	31 %	17 %	95 %

Source: Analytics SME Monitoring Report (2015)

SMEs have to grow and progress to ensure the expected returns on investment and such can be achieved through sustainable superior organizational performance. Many determinants have been identified as influential to organizational performance with few becoming increasingly vital such as knowledge management, innovation and emerging technologies adoption and application (Nawab et al., 2015) which are part of the focus of the paper.

Technology use in workplaces enables the workforce to share resources and communicate efficiently and effectively to encourage innovative ideas. Berisha-Namani (Berisha-Namani, Mihane 2009) states that in the current scenario, use of technology can provide flexibility, better interaction and interconnectivity with partners, customers, suppliers and the market both at national and international levels. It is becoming even more so as societies are now moving from an information society to a high value information or 'knowledge society'. Many organizations have spent substantial amount on the development of technology infrastructure while technology orientation within SMEs is still not as common as compared to large and international enterprises. Ashrafi and Murtaza (Ashrafi, Rafi, and Muhammed Murtaza 2008; Ashrafi, Rafi, and

Muhammed Murtaza 2010; Ashrafi, Rafi, Sujeet Kumar Sharma, Ali H. Al-Badi, and Khamis Al-Gharbi 2014) reported that one of the main barriers in technology adoption is lack of awareness resulting to many countries spending more on technological solutions and awareness programs in order to gain benefits. SMEs have been fairly slow in taking up technology adoption (Jutla, Dawn, Peter Bodorik, and Jasbir Dhaliwal 2002; Lawson, Robyn, Carole Alcock, Joan Cooper, and Lois Burgess 2003; Lopez-Nicolas, Carolina, and Pedro Soto-Acosta 2010).

Global economics has undergone significant changes where decades ago financial capital and labour were considered as mainstay for organizations' progress. This has now been replaced by intellectual capital and knowledge. Knowledge management facilitates the adaptation to market requirements and enhancing efficiency and innovativeness (Bosua, Rachelle, and Krishna Venkitachalam 2013). A substantial number of studies have shown that large enterprises successfully utilize knowledge management and that there is a positive correlation between knowledge management and enterprise growth and development rates (Zack, Michael, James McKeen, and Satyendra Singh 2009). At the same time, however, studies reveal that small- and medium-sized companies do not fully exploit knowledge management tools (Durst, Susanne, and Ingi Runar Edvardsson 2012; Saqib, Muhammad, Nazim Hussain Baluch, and Zulkifli Mohamed Udin 2017; Saqib, Muhammad, Zulkifli Mohammed Udin, and Nazim Baluch 2017).

Acknowledging the fact that knowledge management concept is described in several inconsistent ways, in this paper where SMEs performance is the main concern, KM is referred to as the dynamic combination of three specific measurable components namely Knowledge Identification & Acquisition, Knowledge Dissemination and Knowledge Application which when they are appropriately used, will result in enhanced organizational performance (Saqib, Muhammad, Nazim Hussain Baluch, and Zulkifli Mohamed Udin 2017; Saqib, Muhammad, Zulkifli Mohammed Udin, and Nazim Baluch 2017; Abdullah, Moha Asri 1999; Asoh, Derek Ajesam, and Salvatore Belardo 2007; Zhou, Haibo, and Lorraine M. Uhlaner 2009; Zhou, Haibo, and Lorraine M. Uhlaner 2009).

Durst, Susanne, and Ingi Runar Edvardsson (2012) in a research study on SMEs realized that though there is some awareness in SMEs on KM, it is not holistically approached. Certain areas of knowledge management like dissemination and implementation are strong whereas the areas of knowledge identification, storage and usage lack adequate understanding, and suggested more research to look at the growth of SMEs from that angle. As per the well-known architecture of KM comprising of people or culture, process and technology, technology is considered a major enabler and vital to KM implementation. Several research scholars suggested that Technology Orientation (TO) is valuable to SMEs. In this paper, its potential effects as moderating role on the relationship between knowledge management and performance of SMEs are explored. Few studies that attempt to create association between them lack empirical and consistent evidence and therefore, a more theoretical and empirical approach is being used to investigate the proposed relationship between knowledge management, technology orientation and SMEs performance.

III. Research Methodology

This study aims to use an important and most suitable method called as survey method. Agreeing to Zikmund survey technique (Zikmund, William G., Barry J. Babin, Jon C. Carr, and Mitch Griffin 2013), it is more functional in exploring and development of an idea and also in trying to find the explanations of any specific accomplishment.

The method is well-liked and is very frequently used for conducting quantitative research in the field of management sciences (Bennett, Roger 2003). Thus, based on the literature facts discussed above, it is determined that survey methodology would be suitable for our research. For better sampling, it is very important to take a note of heterogeneity factor in the target population, variables of research and proposed statistical for the data analysis as stated by (Johnson, Burke 2001; Husse, J., and Roger Hussey 1997). Keeping in view the above mentioned considerations, a sample size is calculated based on the table provided by Krejcie and Morgan. Manufacturing SMEs, Sultanate of Oman were segregated / stratified on the basis of "number of SMEs". This study has used "proportionate stratified random sampling" technique to select the sample. Proportionate sampling is suitable as when it is "easier, simpler, and less expensive" to collect data from one or more stratum (Krejcie, Robert V., and Daryle W. Morgan 1970; Sekaran, Uma, and Roger Bougie 2016). The sample was randomly selected according to the population size in each stratum, that is, city in Sultanate of Oman.

Currently, there are three major cities with maximum number of manufacturing industries, that is, Muscat, Sohar and Nizwa. Accordingly sample was collected as shown below.

Table 2:
Data Collection Via Questionnaires Distribution

SNo	City – Strata	Total population	Sample Obtained
1	Muscat	~150	50
2	Sohar	~100	30
3	Nizwa	~50	20
	Total	300	100

(By applying Stratified proportionate Random Sampling)

IV. DEVELOPMENT OF HYPOTHESIS

In conducting a quantitative research, one of the important considerations is hypotheses development. Following (Noble, Charles H., and Minu Kumar 2010), hypothesis is defined as "a proposition that is stated in a testable form and that predicts a particular relationship between two (or more) variables."

The present research study proposed that there is positive relationship between knowledge management and performance of SME in terms of the KM practices that organizations are practicing. The above mentioned hypothesis postulates the relationship between the dimensions (processes) of knowledge management and SMEs performance measures. This means that the SMEs performance improves when the degree of knowledge management processes are practiced. Consequently, in order to investigate the degree of relationship between KM and SMEs Performance, following relationship is hypothesized.

H1: There is a positive relationship between Knowledge Management and Performance of SMEs.

A deep and thorough understanding of TO in terms of SMEs is important not only for academic purposes but also because the subject has salience for practitioners and policy-makers. These implications relate to the profitability and competitiveness of the firm as well as to the overall economic performance of industry and the national economy (Dorf, Richard C., and Thomas H. Byers 2004). The challenge and opportunity for firms in emerging economies is to adopt technology and act entrepreneurially, which can optimize and maximize developmental efforts. Thus, there is a need to study the Technology orientation in a knowledge base and R&D environments (Urban, Boris, and Jose Barreria 2010).

Keeping in view aforementioned discussion, hence, study posits following hypothesis as:

H2: There is a positive relationship between Technology Orientation and Performance of SMEs.

Barron and Kenny (1986), and Kruger & Johnson (2013) posited that in the such cases where relationships between the independent, the predictor variable and dependent, outcome variables are not fully established as significant or insignificant, consistent or inconsistent, in those situations, it is recommended to apply moderator variable to assess its association on the aforementioned relationships.

H3: Technology Orientation moderates the relationship between Knowledge Management and Performance of SMEs.

V. Research Framework, Analysis and Results

The result analysis starts with the presentations of how the individual components of the conceptual framework (see below) were measured, followed by the interpretations of their inter-relationships resulted from the analysis of the collected data.

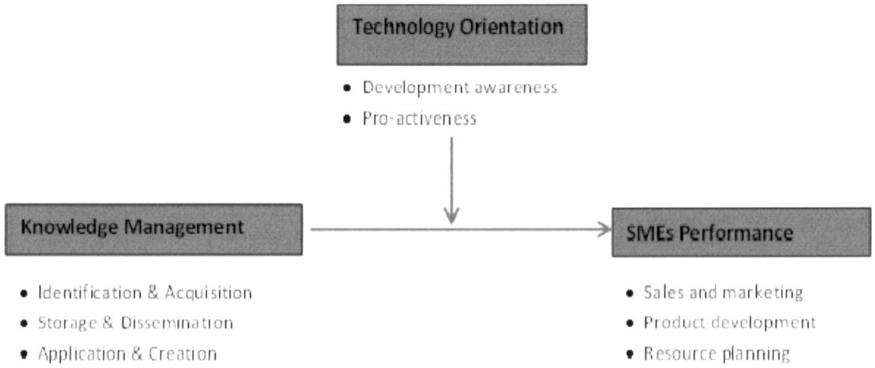

Figure1: Conceptual framework with their measuring criteria

VI. CONTRIBUTING ELEMENTS OF EACH COMPONENT

The contributing elements refer to the three main components showed in the conceptual diagram namely Knowledge Management, Technology Orientation and SMEs Performance.

A. *Knowledge management*

The contributions of knowledge management within the SMEs in Oman are defined by three sub categories namely knowledge identification & acquisition, knowledge storage & dissemination and knowledge application & creation.

Knowledge identification & acquisition relates to elements such as the extent to which new knowledge is achieved through internal and external source tacitly or explicitly, how much of such knowledge is converted into useful forms for further product development, and the enhancing of existing knowledge through the recruitment process. Knowledge identification & acquisition is given an average rating of 3.38 out of 5 showing that manufacturing SMEs in Oman do pay attention to identifying and acquiring new knowledge to further enhance their manufacturing capabilities. Besides the existing workforce, it also shows they integrate the same within their recruitment process to ensure that some sort of knowledge value adding results from the process.

Knowledge storage & dissemination relates to elements such as the extent gained knowledge is properly organized and stored through the use of organizational formal repositories. In other words, converting individualized knowledge into organizational knowledge which can easily be accessed or shared to appropriate targeted colleagues using different means and mechanisms. It further promotes the knowledge sharing culture which mitigates knowledge gaps within organizations due to employee mobility. Knowledge storage & dissemination is given an average rating of 3.53 out of 5 which shows that manufacturing SMEs in Oman do invest effort in building organizational knowledge by converting individualized knowledge to systemic knowledge which later become accessible to appropriate work groups positively impacting the efficiency and effectiveness of the SMEs.

Knowledge application & creation relates to elements such as the extent to which policies and procedures are adhered to so as to protect organizational knowledge while promoting its application to enhance the operational capabilities of individuals, organizations and communities. It also looks at the extent to which feedbacks received from experts are integrated within current practices to improve development processes and product development to further drive the competitive edge. Furthermore, how actively and strategically knowledge creation is incentivized and managed while simultaneously instilling the appreciation in employees that knowledge created while undertaking organizational activities is part of organizational assets. Knowledge application & creation is given an average rating of 3.97 out of 5 which shows that manufacturing SMEs in Oman do emphasize on the adherence to policies and procedures developed through the incorporation of good practices and expert opinions. This ensures that knowledge is well managed and applied during process development and product development.

The analysis gives an average rating of 3.62 out of 5 for knowledge management in general showing that the practice is fairly well embraced by manufacturing SMEs in Oman.

B. ***Technology Orientation***

The contributions of technology within the SMEs in Oman are defined by two sub categories namely technology development awareness and technology pro-activeness.

Technology development awareness relates to elements such as the extent sophisticated technologies are integrated and used across operational units to ensure state of the art product development to keep ahead of competitors. Technology development awareness is given an average rating of 3.88 out of 5 which shows that manufacturing SMEs in Oman well integrate and use sophisticated technologies across their operational units to ensure high quality product development thus being market competitive.

Technology pro-activeness relates to elements such as the extent to which organizations emphasize on R&D with regards to ambitiously creating technical knowledge among individuals and across the organization to develop and maintain better industrial and manufacturing techniques than that of competitors. It also involves the extent to which organizations build relationships with technology suppliers that can lead to customized equipment creating the desirable competitive advantage while at the same time, pro-actively working on internal technology developments and marketing successes which can also provide competitive advantage with regards to industrial standing. Technology pro-activeness is given an average rating of 4.02 out of 5 which shows that manufacturing SMEs in Oman are doing very well in terms of R&D (internal and external) to enhance individual and organizational knowledge relating to keeping up with the development of better industrial and manufacturing techniques and also creating appropriate alliances with suppliers to achieve competitive advantages through the use of customized equipment.

The analysis gives an average rating of 3.95 out of 5 for technology orientation in general showing that technologies are even better embraced than knowledge management practice by manufacturing SMEs in Oman.

C. SMEs performance

Performance is the dependent component of the conceptual framework and it is based on three sub categories (action strategies) resulting from the knowledge management practice and technology orientation namely sales and marketing, product development and resource planning.

Sales and marketing relates to elements such as the extent to which organizations know their market and how it is helping them with their market development strategy and gaining shares, and improving their sales growth rate. Sales and marketing is given an average rating of 4.11 out of 5 which shows that manufacturing SMEs in Oman have a very good understanding of the market they are operating within and that is helping them in developing their marketing strategies and expanding their market share.

Product development relates to elements such as the extent to which organizations continuously undertake research activities that result in innovative product development. Product development is given an average rating of 4.15 out of 5 which shows that manufacturing SMEs in Oman are continuously finding ways through research activities to be more innovative when it comes to product development.

Resource planning relates to elements such as the extent to which organizations appropriately invest in organizational resources be it human resources or not, to optimize operating benefits thus return on investment. Resource planning is given an average rating of 4.18 out of 5 which shows that manufacturing SMEs in Oman appropriately invest in resources required to achieve optimal operating performance to ensure return on investment.

The analysis gives an average rating of 4.15 out of 5 for performance in general showing that manufacturing SMEs in Oman feel that they are achieving very good performance.

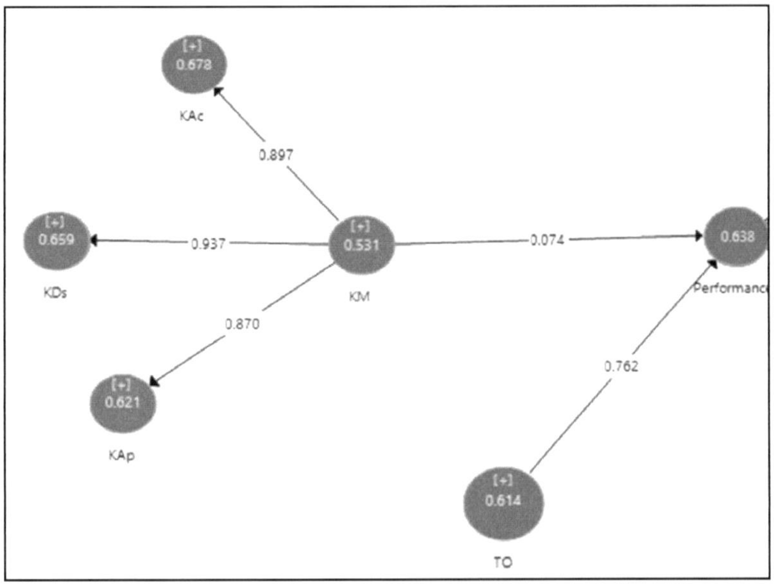

Figure2: KM-TO-Performance Framework using Smart PLS (Researcher)

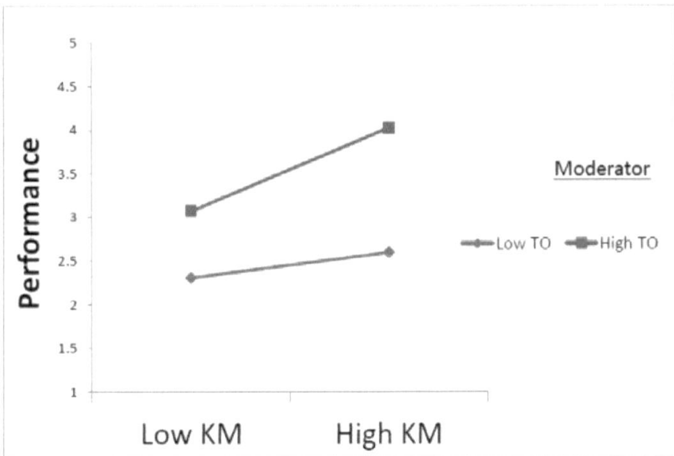

Figure3: Moderating Effect of TO on KM-Performance Relationship (Researcher)

It can be noted that knowledge identification & acquisition, knowledge storage & dissemination and knowledge application & creation are fairly well practiced within the manufacturing SMEs in Oman with individual practice level of 0.678, 0.659 and 0.621 respectively. They significantly influence the knowledge management practice within the manufacturing SMEs in Oman with correlating values of 0.897, 0.937 and 0.870 respectively. Yet still, the overall measurement for knowledge management practice within the manufacturing SMEs in Oman is only 0.531 showing that knowledge management practice is still fairly limited. The components of knowledge management seemed to be operating as isolated activities rather than together as a process. Consequentially, the level of knowledge management being practiced within the manufacturing SMEs in Oman is having minimal influence on the organizational performance with a correlating value of only 0.074. But interestingly, the intervention of technology as a moderating variable which is also fairly well practiced with a level of 0.614 within the manufacturing SMEs improves the influencing factor significantly with a correlating value of 0.762 making a reasonable organizational performance of 0.638 achievable. It would seem that currently the perceptions are fragmented (Durst, Susanne, and Ingi Runar Edvardsson 2012) but if the components of knowledge management could be better combined into synchronized processes within the manufacturing SMEs in Oman, better correlating values between knowledge management and organizational performance could be achieved thus further improvement in overall organizational performance. At present, the moderating variable which is technology is playing a crucial role on the overall organizational performance as shown on the moderating effect graph below. This works well for the manufacturing SMEs in Oman due to the high penetration of technology within that industry (Ashrafi, Rafi, and Muhammad Murtaza 2010). From the graph, it can be clearly noted that the growth in the overall organizational performance increases at a greater rate as the intervention of technology increases.

Table 3:
Hypotheses with Decision

Hypotheses	**Relationship**	**Full Model**			**Decision**
		Std. Beta	***Std. Error***	***t-Value***	
H1	Knowledge Management -> SME Performance	0.074	0.07	2.396**	Supported
H3	Technology Orientation -> SME Performance	0.762	0.06	2.421**	Supported
H4	Tech-Orientation* Knowledge Management -> SME Per	0.166	0.102	3.065**	Supported

**p<0.01 *T values >1.64 (p values < 0.01)*

VII. RECOMMENDATIONS:

Recommendation section highlights a number of activities that can be considered to encourage such practices.

1. **Emerging capability for and supporting research & development in SMEs sector (R&D – SMEs Sector):** Innovation relates to application/transmission of knowledge from research to development of products/services. Innovative organizations are strongly research & development (R&D) oriented and are proactive in acquiring novel information while using sophisticated technologies (Cooper,1984; Kanter, 1988). Little emphasis is currently being placed on R&D in manufacturing SMEs in Oman neither to acquire novel knowledge nor to embrace the continuous development of good practices within the sector. This is mainly due to limited research capabilities and the culture of non-research based decision making. Building research capabilities of SMEs and changing the culture regarding research is an effective means for a long term solution as it would require significant time and effort while a more immediate solution is to develop a working relationship between the SMEs sector and the bodies/authorities that already have the research capabilities. For instance, the Oman Research Council which is a Government affiliated body can work alongside the SMEs to undertake the more pressing research needs while simultaneously providing some guidance to them on how to integrate basic research and innovation within their current practices. The outcomes of the researches can then be made accessible to all SMEs to promote good practices which can in turn boost their performances and growth.

2. **Encouraging collaboration among SMEs**: Collaborative innovation is becoming increasingly important as it promotes the knowledge sharing culture which normally leads to good practices thus resulting in overall performance enhancement. Currently, minimal or no collaboration is formally being practiced in between SMEs and therefore diffusion of good practices and newly created knowledge through the SME sector is very limited (Rocha, 2017). A knowledge management initiative supported by technology where

good practices may be exchanged or shared among SMEs and new knowledge may be generated would be helpful. Such initiatives are becoming very common in SMEs sector and proven to be very successful in countries such as Thailand, Bangladesh, Pakistan, China, Singapore, and Malaysia and so on. Capabilities and resources required to implement such initiative are likely to be beyond the SMEs in Oman and therefore the support will have to come from the Government which is also well positioned to incentivize the SMEs to participate.

3. **Revising of Government legislation and support towards SMEs**: Challenges faced by most developing countries' SMEs have restricted access to top quality raw materials and lack of experienced human resource to take advantage of information and communication technologies for business practice improvements (Mutula, Stephen M., and Pieter van Brakel 2006). The absence of knowledge management and innovation culture, insufficient marketing and managerial skills, and difficulty in accessing easy financing are add on challenges (Qiang, Christine Zhen-Wei, George R. Clarke, and Naomi Halewood 2006; Qiu Zhang, Hanqin, and Alison Morrison 2007). Furthermore, SMEs in developing countries tend to have lesser market shares due to their less than ideal infrastructure and their struggles to integrate regional cooperation, governance, legal and administrative processes. (Ongori, Henry 2008). Oman being a developing country, its SMEs are facing similar issues and therefore further commitment and support is required from the Government. Improved or new trade deals have to be put in place that would allow SMEs to have easy access to and affordable importation of high quality raw materials. This should be accompanied by accessible favorable financing to ensure that SMEs could actually make use of the deals. As a result, the quality of the products being produced by Omani SMEs could be greatly enhanced thus becoming more marketable beyond the Omani border. With the right marketing strategies through the same trade deals, the potential export market size and reach could significantly increase and so would be the growth of the SMEs sector. As explained by the Central Bank of Oman, reduction in rigidities and some push in the manufacturing sector could improve competitiveness and performances in the Sultanate (CBO, 2016).

4. **Striving for a more competitive market**: Currently, in Oman, market trends or behaviors especially in SMEs tend to be quite stagnant as there is not much of a competition within the market (Al-Shanfari, Dhafir Awadh 2012). SMEs are operating quite comfortably and safely within the Omani market to the extent where the need to grow and enhance the quality of existing products/services or even consider developing new ones does not come high on their agenda. But as discussed earlier, SMEs have a major dynamic role to play within a developing economy and it should be the same case for Oman where the SMEs sector is pretty static (CBO, 2016). Therefore, the need to push for a more competitive market should get the SMEs to become more active regarding the economic development.

VIII. LIMITATIONS AND FUTURE RESEARCH DIRECTIONS

This study has been able to provide support for the hypothesized paths between the predictors and the criterion variable; however, the results also consider some limitations in the study.

First, this study was carried out using a cross-sectional survey design to collect data at a single-point-in-time, which prevents causal inferences to be made from the population of the study. Therefore, it is suggested that an alternative research design, like longitudinal design, should be considered in the future studies, as this will allow the latent construct's measurement at different points in time in order to acquire more understanding of this study's model.

Secondly, the present study was conducted among Omani small and medium scale manufacturing firms only. In this way, it offers a limited generalizability. Although, studies have noted that knowledge management implementing SMEs are known to perform better in several processes like return on investment, customers satisfaction, employee development, and market share and development (Anantatmula, Vittal, and Shivraj Kanungo 2006), it should be noted that SMEs are constrained due to their size and resource inadequacy unlike larger firms (Sullivan Mort, Gillian, Jay Weerawardena, and Peter Liesch 2012). However, other previous studies have revealed that technology-oriented SME firms are usually characterized by improved performance like R&D activities, which strengthens their technological base. Therefore, future studies should extend this study to include larger manufacturing firms who have shown to be more responsive to knowledge management, innovate quickly and more technology-oriented in their organization's overall mission than their SME counterparts (Brett, Valerie, Bill O'Gorman, and Óscar Afonso 2017; Byukusenge, Eugenie, and John C. Munene 2017).

IX. CONCLUSION

This present study has provided additional contribution to the body of knowledge on the moderating effects of technology orientation on the relationship between knowledge management, and SMEs performance. The findings of this study have provided empirical support to all of the research questions formulated despite few limitations. As regards the five hypothesis statements highlighted earlier in this study, three hypotheses (two direct and one moderating effect) were statistically significant. While there have been several studies that examined several underlying factors influencing firm's performance among manufacturing SMEs, this study has addressed a theoretical gap by incorporating technology orientation as a significant moderating variable. In the same manner, this study also provided theoretical and empirical support for the moderating role of technology orientation on the relationship between knowledge management and SMEs performance. The framework developed in this study has contributed to the domain of the theory of growth of firm and resource-based view theory, and by extension, to the fields of knowledge-based view and dynamic capabilities perspective, by examining the influence of knowledge management and technology orientation on SMEs performance.

The notion of this research is that knowledge management and technology have an impact on the performance of organizations within this emerging knowledge-economy era. Many organizations depend upon knowledge and have untied the research prospects to fill the existing gaps.

To promote knowledge management within SMEs in Oman to the extent that it can provide that expected positive thrust to their performances, a number of activities have been identified as possible catalyst. The first one being the developing of capability for or supporting research & development in SMEs sector, the second being the promoting of collaboration among SMEs, the third being the reviewing of Government legislation and support towards the SMEs, the fourth being having a well-focused capacity building plan (Human resource development plan), and finally being the striving for a more competitive market. Many barriers to adopt emerging technologies and factors that are associated to limited adopted technology in SMEs firms. The challenge thus lies in dealing with the implications of technology within organizations, the effects of which may lead to a positive change in the organization. Technology orientation may further solve the impending problems, reduced cost and strengthen the overall performance as also shown in the above stated empirical research.

Consequently, it will ensure better understanding of the business environments, save time-to-markets (reduce product/process development life cycle time), promote learning and enhance problem solving skills, and encourage integration of new knowledge to stimulate creative products and services. Consistent with these findings, policy makers need to emphasize more on knowledge and technology as important resources in the manufacturing strategy processes and manufacturing performance.

REFERENCES

Brief, OECD Policy. "Small and Medium Sized Enterprises: Local Strength Global Reach." (2000)

Shackelford, Adam, and Peter Yih-Tong Sun. "Knowledge management strategy employed in a SME: The case of a building materials supplier." (2009): 1-18.

Sandberg, Karl W. "An exploratory study of women in micro enterprises: gender-related differences." *Journal of small business and enterprise development* 10, no. 4 (2003): 408-417.

Verheugen, Günter. "The new sme definition: user guide and model declaration." *Enterprise and Industry Publications, European Commission* (2005).

Dollinger, Marc J. "Environmental boundary spanning and information processing effects on organizational performance." *Academy of Management Journal* 27, no. 2 (1984): 351-368.

Dollinger, Marc J. "Environmental contacts and financial performance of the small firm." *Journal of Small Business Management (pre-1986)* 23, no. 000001 (1985): 24.

Brush, Candida G., and Pieter A. Vanderwerf. "A comparison of methods and sources for obtaining estimates of new venture performance." *Journal of Business venturing* 7, no. 2 (1992): 157-170.

Davenport, Thomas H., and Laurence Prusak. *Working knowledge: How organizations manage what they know.* Harvard Business Press, 1998.

Egbu, Charles O., Subashini Hari, and Suresh H. Renukappa. "Knowledge management for sustainable competitiveness in small and medium surveying practices." *Structural survey* 23, no. 1 (2005): 7-21.

Sabri, Hala. "Knowledge management in its context: adapting structure to a knowledge creating culture." *International Journal of Commerce and Management* 15, no. 2 (2005): 113-128.

Daud, Salina, Rabiah Eladwiah Abdul Rahim, and Rusnita Alimun. "Knowledge creation and innovation in classroom." *International Journal of Human and Social Sciences* 3, no. 1 (2008): 75-79.

Murthy, Y. Sree Rama. "Managing Small and Medium Enterprises in Oman and Financing Problems." *i-Manager's Journal on Management* 4, no. 2 (2009): 68.

Belwal, Shweta, Rakesh Belwal, and Fatema Al Saidi. "Characteristics, motivations, and challenges of women entrepreneurs in Oman's Al-Dhahira Region." *Journal of Middle East Women's Studies* 10, no. 2 (2014): 135-151.

Atef, Tamer Mohamed, and Masooma Al-Balushi. "Entrepreneurship as a means for restructuring employment patterns." *Tourism and Hospitality Research* 15, no. 2 (2015): 73-90.

Bilal, Zaroug Osman, and Nawal Said Al Mqbali. "Challenges and constrains faced by small and medium enterprises (SMEs) in Al Batinah governorate of Oman." *World Journal of Entrepreneurship, Management and Sustainable Development*11, no. 2 (2015): 120-130.

McWilliam, Tyler Philip Eldridge. "Made in Oman: Promoting Manufacturing and Export in the Sultanate of Oman." (2011).

Thorpe, Richard, Robin Holt, Allan Macpherson, and Luke Pittaway. "Using knowledge within small and medium- sized firms: a systematic review of the evidence." *International Journal of Management Reviews* 7, no. 4 (2005): 257-281.

Nawab, Samina, Tahira Nazir, Muhammad Mohsin Zahid, and Syed Muhammad Fawad. "Knowledge management, innovation and organizational performance." *International Journal of Knowledge Engineering-IACSIT* 1, no. 1 (2015): 43-48.

Berisha-Namani, Mihane. "The role of information technology in small and medium sized enterprises in Kosova." In Fulbright Academy Conference, pp. 1-8. 2009.

Ashrafi, Rafi, and Muhammed Murtaza. "Use and impact of ICT on SMEs in Oman." *Electronic Journal of Information Systems Evaluation* 11, no. 3 (2008).

Ashrafi, Rafi, and Muhammad Murtaza. "ICT adoption in SME in an Arab GCC Country Oman." *E-strategies for resource management systems: Planning and implementation* (2010): 371-376.

Ashrafi, Rafi, Sujeet Kumar Sharma, Ali H. Al-Badi, and Khamis Al-Gharbi. "Achieving business success through information and communication technologies adoption by small and medium enterprises in Oman." *Middle-East Journal of Scientific Research* 22, no. 1 (2014): 138-146.

Jutla, Dawn, Peter Bodorik, and Jasbir Dhaliwal. "Supporting the e-business readiness of small and medium-sized enterprises: approaches and metrics." *Internet Research* 12, no. 2 (2002): 139-164.

Lawson, Robyn, Carole Alcock, Joan Cooper, and Lois Burgess. "Factors affecting adoption of electronic commerce technologies by SMEs: an Australian study." *Journal of small business and enterprise development* 10, no. 3 (2003): 265-276.

Lopez-Nicolas, Carolina, and Pedro Soto-Acosta. "Analyzing ICT adoption and use effects on knowledge creation: An empirical investigation in SMEs." *International Journal of Information Management* 30, no. 6 (2010): 521-528.

Bosua, Rachelle, and Krishna Venkitachalam. "Aligning strategies and processes in knowledge management: a framework." *Journal of Knowledge Management* 17, no. 3 (2013): 331-346.

Zack, Michael, James McKeen, and Satyendra Singh. "Knowledge management and organizational performance: an exploratory analysis." *Journal of knowledge management* 13, no. 6 (2009): 392-409.

Durst, Susanne, and Ingi Runar Edvardsson. "Knowledge management in SMEs: a literature review." *Journal of Knowledge Management* 16, no. 6 (2012): 879-903.

Saqib, Muhammad, Nazim Hussain Baluch, and Zulkifli Mohamed Udin. "Moderating Role of Technology Orientation on the Relationship between Knowledge Management and SMEs' Performance in Oman: A Conceptual Study." International Journal of Economic Perspectives 11, no. 1 (2017): 433-441.

Saqib, Muhammad, Zulkifli Mohammed Udin, and Nazim Baluch. "The Impact of Knowledge Management on Organizational Performance in Today's Economy." South East Journal of Contemporary Business, Economics and Law 12, no. 3 (2017).

Abdullah, Moha Asri. "The accessibility of the government-sponsored support programmes for small and medium-sized enterprises in Penang." *Cities* 16, no. 2 (1999): 83-92.

Asoh, Derek Ajesam, and Salvatore Belardo. "Assessing knowledge management: Refining and cross validating the knowledge management index using structural equation modeling techniques." *International Journal of Knowledge Management (IJKM)* 3, no. 2 (2007): 1-30.

Zhou, Haibo, and Lorraine M. Uhlaner. "Knowledge Management in the SME and its Relationship to Strategy, Family Orientation and Organization Learning." (2009).

Zhou, Haibo, and Lorraine M. Uhlaner. "Knowledge Management as a Strategic Tool to Foster Innovativeness of SMEs." (2009).

Zikmund, William G., Barry J. Babin, Jon C. Carr, and Mitch Griffin. *Business research methods*. Cengage Learning, 2013.

Bennett, Roger. "Factors underlying the inclination to donate to particular types of charity." *International Journal of Nonprofit and Voluntary Sector Marketing* 8, no. 1 (2003): 12-29.

Johnson, Burke. "Toward a new classification of nonexperimental quantitative research." *Educational Researcher* 30, no. 2 (2001): 3-13.

Husse, J., and Roger Hussey. "Business research: a practical guide for undergraduate and postgraduate students." *Basingstock: Macmillan Business* (1997).

Krejcie, Robert V., and Daryle W. Morgan. "Determining sample size for research activities." *Educational and psychological measurement* 30, no. 3 (1970): 607-610.

Sekaran, Uma, and Roger Bougie. *Research methods for business: A skill building approach*. John Wiley & Sons, 2016.

Noble, Charles H., and Minu Kumar. "Exploring the appeal of product design: A grounded, value- based model of key design elements and relationships." *Journal of Product Innovation Management* 27, no. 5 (2010): 640-657.

Dorf, Richard C., and Thomas H. Byers. *MP: Technology Ventures: From Idea to Enterprise w/Engineering Subscription Card*. McGraw-Hill Science/Engineering/Math, 2004.

Urban, Boris, and Jose Barreria. "Empirical investigations into firm technology orientation and entrepreneurial orientation." *International Journal of Innovation and Technology Management* 7, no. 04 (2010): 329-351.

Baron, Reuben M., and David A. Kenny. "The moderator–mediator variable distinction in social psychological research: Conceptual, strategic, and statistical considerations." *Journal of personality and social psychology* 51, no. 6 (1986): 1173.

Kruger, Cornelius Neels, and Roy D. Johnson. "Knowledge management according to organisational size: A South African perspective." *SA Journal of Information Management* 15, no. 1 (2013): 11-pages.

Becerra-Fernandez, Irma, and Rajiv Sabherwal. *Knowledge management: Systems and processes*. Routledge, 2014

Pirnea, Ionela Carmen, and Elena Alexandra Caldararu. "The relationship between performance and knowledge management in Romanian SMEs." *Anale. Seria Stiinte Economice. Timisoara* 18 (2012): 514.

Powell, Walter W., and Kaisa Snellman. "The knowledge economy." *Annu. Rev. Sociol.* 30 (2004): 199-220.

Liao, Shu-hsien, and Chi-chuan Wu. "The relationship among knowledge management, organizational learning, and organizational performance." *International Journal of Business and Management* 4, no. 4 (2009): 64.

RIZEA, Ionela Carmen, Denisa Elena PARPANDEL, Alexandra Elena CĂLDĂRARU, and Andreea Lorena RADU. "Current approaches regarding the knowledge management impact on SMEs performance." *Economia. Seria Management*14, no. 2 (2011): 563-568..

Skyrme, David. *Capitalizing on knowledge*. Routledge, 2012.

Kanter, Rosabeth Moss. "Three tiers for innovation research." *Communication Research* 15, no. 5 (1988): 509-523.

Mutula, Stephen M., and Pieter van Brakel. "E-readiness of SMEs in the ICT sector in Botswana with respect to information access." *The electronic library* 24, no. 3 (2006): 402-417.

Banker, Rajiv D., Abraham Charnes, and William Wager Cooper. "Some models for estimating technical and scale inefficiencies in data envelopment analysis." *Management science* 30, no. 9 (1984): 1078-1092.

Qiang, Christine Zhen-Wei, George R. Clarke, and Naomi Halewood. "Information and communications for development: Global trends and policies." (2006).

Qiu Zhang, Hanqin, and Alison Morrison. "How can the small to medium sized travel agents stay competitive in China's travel service sector?." *International Journal of Contemporary Hospitality Management* 19, no. 4 (2007): 275-285.

Ongori, Henry. "Barriers to ICTs adoption in SMEs: Evidence from a developing country perspective." *Prabandhan: Indian Journal of Management* 1, no. 2 (2008): 12-17.

Rasula, Jelena, Vesna Bosilj Vuksic, and Mojca Indihar Stemberger. "The impact of knowledge management on organisational performance." *Economic and Business Review for Central and South-Eastern Europe* 14, no. 2 (2012): 147.

CBO, *Annual Report of Oman.* Muscat: Cental Bank of Oman (2016).

Al-Shanfari, Dhafir Awadh. "Entrepreneurship in Oman: A snapshot of the main challenges." In *Proceedings of the United Nations Conference on Trade and Development: Multi-year expert meeting on enterprise development policies and capacity-building in science, technology and innovation (STI)*. 2012.

Anantatmula, Vittal, and Shivraj Kanungo. "Structuring the underlying relations among the knowledge management outcomes." *Journal of knowledge management* 10, no. 4 (2006): 25-42.

Sullivan Mort, Gillian, Jay Weerawardena, and Peter Liesch. "Advancing entrepreneurial marketing: Evidence from born global firms." *European Journal of Marketing* 46, no. 3/4 (2012): 542-561.

Brett, Valerie, Bill O'Gorman, and Óscar Afonso. "Supporting Knowledge and Policy-Based Stakeholders in Delivering Regional Impact: A Tool to Select Regional Scoreboard Indicators." In *The Quadruple Innovation Helix Nexus*, pp. 137-171. Palgrave Macmillan, New York, 2017.

King, William R. "Knowledge management and organizational learning." In *Knowledge management and organizational learning*, pp. 3-13. Springer US, 2009.

Byukusenge, Eugenie, and John C. Munene. "Knowledge management and business performance: Does innovation matter?." *Cogent Business & Management* 4, no. 1 (2017): 1368434.

Dalkir, Kimiz. *Knowledge management in theory and practice*. Routledge, 2013.

Hamzehi, A., Salimi, F. and Khodayari, M. (2016) 'The Relationship between Business Intelligence and Knowledge Management Dimensions', Account and Financial Management Journal.

Smart Stick for blind people

Wala Shajanah [a], Noor AL-Shahri [a], , Waad AL-Rubeai [a], Raida Rabia [a]

Supervisor: Dr. Najam ul Hasan [a]

[a] Electrical Engineering Department, College of Engineering, Dhofar University, PO Box 2009, Postal Code 211 Salalah, Sultanate of Oman

nulhasan@du.edu.om

DHOFAR UNIVERSITY

جامعة ظفار Dhofar University

Abstract

Blind people are ordinary people whose visual abilities are different or have disappeared because of genetic or pathological causes or because of their exposure to accidents in different ways. So many people in the world are exposed to blindness that remains trouble throughout their lives. An electronic stick can help tool for these people to live a normal life. In this project, we developed a smart stick with a number of features that can be very helpful for a blind person to walk around easily. This stick has a number of ultrasonic sensors to detect any obstacle close to the person. It has a buzzer to inform him about any dangerous situation. Moreover, it has a push button, by pressing it; the person can inform his caretakers about his position and can call them for help.

PROBLEM STATEMENT

How a stick in the hands of a blind person can be helpful in his life?

BLOCK DIAGRAM

SYSTEM WORKING

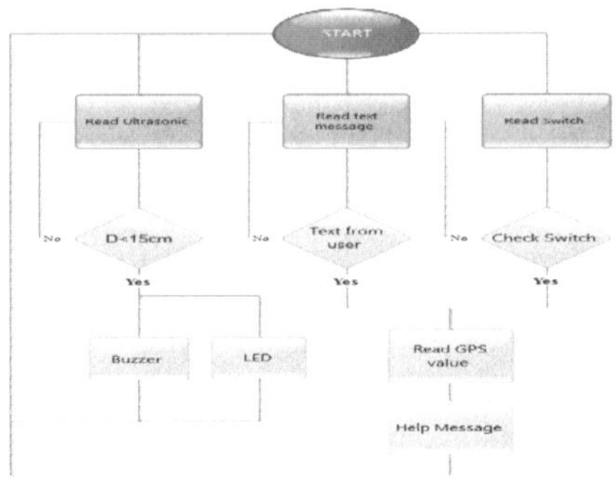

SIMULATION

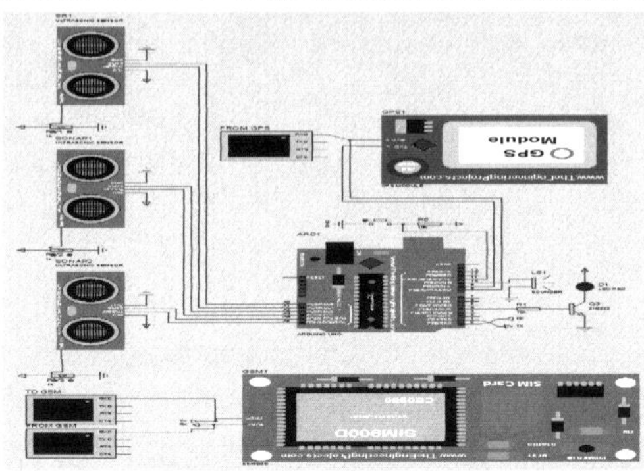

HARDWARE PROTOTYPE

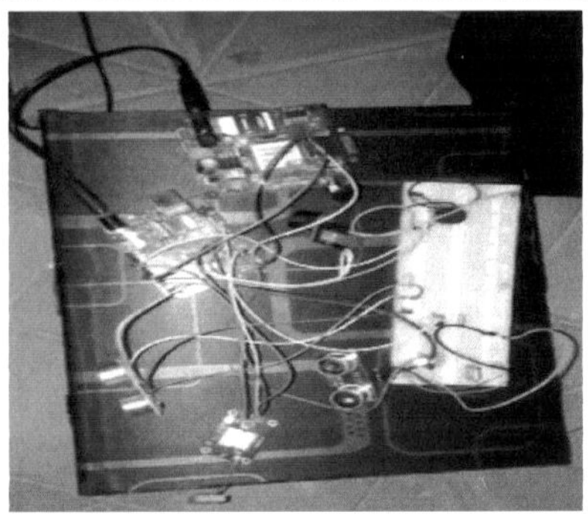

CONCLUSION

A stick is designed and implemented to ease the life of blind persons.

Analysis of the failing E-Commerce industry in Indian Market: Case Study of Flipkart Pvt. Ltd.

S S Rizwanullah Quadri and Sayyada Mehrina Quadri

Lecturer, MGES Khan Abdul Ghaffar Khan College, Pathri, Parbhani, India
Email: quadri.rizwan@gmail.com
Email: mehrinaquadri@gmail.com

This paper is an effort to point out the challenges and opportunities of Indian Ecommerce industry with special regards to Etail (e-retail) segment with Flipkart Pvt. Ltd. (Flipkart) as a case study. The E-commerce industry in India is continuously evolving and developing rapidly as more companies, both local and foreign with a spectrum of portfolios are investing heavily in the Indian market due to, its robust consumer base, recent taxation policy reforms, and lower cost overheads in infiltration and Foreign Direct Investment (FDI) benefits. At its current growth rate, the E-commerce industry is expected to grow exponentially by USD 1.3 trillion by 2020 with a current etail of USD 15 billion. Established Economists opine that Indian E-Commerce market shall continue to expand further in many domains. This resultantly will lead to a favorable impact on the Indian economy and consequently aid the Gross Domestic Product (GDP) elevation. However, in contrast, leading economic think tanks of international repute based on the macro and micro analysis of the trends and governance models of various E-Commerce companies, have hinted that the sustainability of such establishments is a major concern in the medium and the long run which stands at an encore with failure. The major hindrances that Flipkart is faced with in its current scenario is not only to deal with the market competition and providing robust deals with huge discounts that are orchestrated to acquire and retain customer loyalty, but also the internal factors such as setbacks in Search Engine Optimization (SEO), Gross Merchandise Volume (GMV) and the overheads in 'warehouse based inventory model' are vital and acts as decisive factors. Furthermore, the demonetization and GST created volatility in the market and an overall slowdown was experienced in all walks of business activities and as well as the regression spiked due to the lesser internet market penetration that stands to 17 % and not covering tier 2 and tier 3 cities. The study uses Sample survey method with a volume of 200 samples to achieve the objectives mentioned supra and 10 percent of the gathered sample survey is randomly used as the primary data in the form of questionnaire. The data is analyzed by SPSS for 2015-2017 the time span that has been turbulent in the etail industry even for Flipkart since its inception. Experts on the other hand believe that for companies like Flipkart, one of the leading e-commerce companies in India, going through the consolidation phase is a necessary step in order to attain sustainable growth in the future. The inferences conclude that although Flipkart revenue has grown by 29 percent (USD 3.09 billion) compared to previous year exponentially, so has its losses elevated with 68 percent (USD 814 million) and a significant proportion can be attributed to the total migration on app and warehouse model system of the inventory till it is delivered to the end customer contributing to the systemic failure of etail industry Flipkart unless a workaround is consolidated in time.

Keywords: E-Commerce, Etailng, Business Sustainability, Overhead Vulnerability, SEO and GMV, Consolidation

References:

Keith and Gupta (2015). Analysis of the Expansion of E commerce Into India and Growth Opportunities for Flipkart, *Journal of International Business and Law*, 14(1), 151.

Kumar, A. (2017) Present and Future of E-Retailing in India: A Case Study of Flipkart, *International Journal of Recent Trends in Engineering & Research* , 3(4), 2455-1457

Ladda, S., 2015, *eCommerce in India Accelerating growth,* Technology Sector Practice, PwC India.

Nigam, P. Nayak, K., Vyas, P.(2015) E- Commerce Challenges: A Case Study of Flipkart. Com versus Amazon. In, *Indian Journal of Applied Research,* 5(2), 2249-555X

Rawat, D. Sharma, M., 2014, *Evolution of e-commerce in India Creating the bricks behind the clicks*, Corporate Communications, PWC India

Sudha, S., Bharathi, K.(2015) Case Study on Flipkart, *International Journal of Transformations in Business Management*, 5(4), 2231-6868.

CXOtoday News Desk, 2016. *E Commerce Fueling India's Employment Generation: Report.* [Online] Available at: http://www.cxotoday.com/story/e-commerce-sector-fueling-employment-generation-in-india-report/ [Accessed 27 December 2017].

De, S., 2017. *E-Commerce – Sustainable business model?.* [Online] Available at: http://www.groupdiscussionideas.com/e-commerce-sustainable-business-model/ [Accessed 27 December 2017].

Khurana, A., 2017. *Learn the Basics About Ecommerce,* [Online] Available at: https://www.thebalance.com/let-s-ask-the-basic-question-what-is-ecommerce-1141599 [Accessed 09 February 2018].

Preethi, J., 2011. *Flipkart Acquires Digital Content From Bollywood News Site Chakpak.* [Online] Available at: http://techcircle.vccircle.com/500/flipkart-acquires-bollywood-news-site-chakpak-com/ [Accessed 27 December 2017].

A Gulf College Case Study: Impact of Bilingual Education Programs on Arabic Language Learners

Rowena Ocier Sibayan, Ed. D.
Gulf College, Oman
rowena.sibayan@gulfcollege.edu.om

Ahlam Ibrahim Al Said Shukaili
Gulf College, Oman
ahlam@gulfcollege.edu.om

Norman S. De Ocampo
Al Seeb Vocational College
namron32610@gmail.com

Abstract

One of the major challenges facing educators and policymakers today in the Arab country is the growing number of limited English proficiency. The main goal of every learning institution is to prepare the students for the 21st century education. Many learning institutions that teach a language that were attended by many students who are not proficient. Learning with communication barriers in regular classrooms put students at risk because of poor performance. To produce competitive learners, all learning institutions must align the curriculum to 21st century education to meet the demand of the global market.

The aim of this study is to explore the impact of bilingual education (BE) on Arab language learners to compete to the demand of 21st century skills in the Sultanate of Oman, but all over the globe. According to the study from the United States Census Bureau (2011), a professional and successful person needs to be bilingual in the 21st century, where English is one of the two languages because it is a universal language in any fields in the world. In the Sultanate of Oman, the highest number of speaker is in English (particularly expatriates) after Arabic. Different agencies including the Ministry of Education and the Ministry of High Education proposed variety of education programs to improve students' performance to produce globally competitive graduates. Moreover, this goal can be achieved with the proper guidance from the parents, educators, and supports from the education stakeholders by providing the appropriate learning resources, materials, and infrastructure. Like other Arab countries, the Sultanate of Oman has many limited English proficient students and some institutions are using bilingual education. This study focuses on the impact of bilingual education on Gulf College Students' performances. The researcher used both qualitative and quantitative research methodology; however, this study focuses more on the qualitative methodology.

The researchers selected the modules based on high number of non-submission as the basis of study, as well as the willingness of the teachers to initiate, conduct teaching lessons for this study. The modules are Mathematics and Technical English for Computing. Researchers identified three phases to evaluate the impact of bilingual education lessons that will be conducted by only one teacher to get the desired output. The first phase of the study was to conduct a teaching lesson in English and assessed the students' learning outcomes. The second phase is to conducted the same lesson in Arabic and assessed the student's learning outcomes with the same teacher in phase 1. In the last phase, the same lesson was conducted by the same teacher both in English and Arabic language and assessed the student's learning outcomes. Students' achievements within these processes were measured by looking at the achievement gap to which the aim is to narrow the gap. The proposed outcome of this study is to have a well-designed bilingual education programs that can have a significant effect on student achievement both in all academic core modules and English literacy.

Keywords: *education, bilingual education, 21st century learner, impact, curriculum*

References

Al Balushi, K., Al Rahbi, H. & Sibayan, R. (2017). "*Oman Education System: The Influence of Social Media in Learner's Learning Process*". AOUIRC 2017, Muscat

Al-Khatib,M. (2000). *The Arab world: Language and cultural issues. Language, Culture, &Curriculum*, 13(2), 121-125

Al-Nakhalah Ahmed (2006). *Tenses Difficulties That Face Students of English in Palestinian Universities While translating from English to Arabic- A Case Study of Al-Quds Open University*

Al Nofaie, H (2010) "*The attitudes of teachers and students towards using Arabic in EFL classrooms in Saudi public schools - a case study*". Novitas-royal (research on youth and language)", 4 (1), 64-95.

Al Shaibany, Saleh (2016). *Oman Education: Bilingual schools prepare children better for top careers.* Times of Oman, Feb 27, 2016. [online] Available at http://timesofoman.com/article/78336. Accessed 5, Jan 2018.

Aquino, L. (2012). *The Effects of Bilingual Instruction on the Literacy Skills of Young Learners | ELTWO*. [online] Blog.nus.edu.sg. Available at: http://blog.nus.edu.sg/eltwo/2012/06/14/the-effects-of- bilingual-instruction-on-the-literacy-skills-of-young-learners-2/ [Accessed 15 Feb. 2018].

Association of American Colleges and Universities (2008). *College Learning for the New Global Century.* Washington, DC. Author.

Cariman, G. (2015). [online] Unesco.org. Available at: http://www.unesco.org/fileadmin/Multimedia/Field/Santiago/pdf/Apunte09-ing.pdf [Accessed 15 Feb. 2018].
Conley, D.T. (2007). Thrive. *The Skills Imperative.* Washington, DC.
Ed.gov. (2016). *The Importance of Bilingual Education | U.S. Department of Education*. Available at: https://www.ed.gov/news/speeches/importance-bilingual-education [Accessed 4 Jan. 2018].
Golden, Claudia. and Katz, Lawrence F. (2008). *The Race Between Education and Technology.* Boston: Harvard University Press.
Healthofchildren.com. (2018). *Bilingualism/Bilingual Education - stages, Definition, Description, Common problems.* Available at: http://www.healthofchildren.com/B/Bilingualism-Bilingual- Education.html#ixzz53QsSMFYq [Accessed 3 Jan. 2018].
Kirkwood, A. & Price, L. (2005). *Learners and learning in the twenty-first century: What do we know about students' attitudes towards and experiences of information and communication technologies that will help us design courses?* Studies in Higher Education 30(3), 257-274
Knoema. (2018). *Oman Adult literacy rate, 1970-2017*. Available at: https://knoema.com/atlas/Oman/topics/Education/Literacy/Adult-literacy-rate [Accessed 5 Jan. 2018].
Learn.org. (2003). *What Is Bilingual Education*?. [online] Available at: https://learn.org/articles/What_is_Bilingual_Education.html [Accessed 7 Jan. 2018].
Mohammad Shaheen, (1990)-*Theories of Translation and Their Application of the Teaching of English/Arabic*, Dar Al-Thaqafa Library.
Mourtaga, K. (2004). "*Investigating writing problems among Palestinian students: Studying English as a foreign language*". Bloomington, Indiana, Author House. (UMI No. 3159503).
Salha A Issan, S. (2018). *Cite a Website - Cite This For Me*. [online] Periglobal.org. Available at: http://www.periglobal.org/sites/periglobal.org/files/The%20period%20of%20change%20Privatization%20Oman.pdf [Accessed 15 Feb. 2018].
Schaub, M. (2000). "*English in the Arab Republic of Egypt*". World Englishers 19, 225- 238. Sde.ct.gov. (2011). Capell. [online] Available at: http://www.sde.ct.gov/sde/lib/sde/pdf/curriculum/bilingual/CAPELL_SPED_resource_guide.pdf [Accessed 15 Feb. 2018].
Sibayan, R. Halibas, A. & Sibayan, K. (2017). "*A Gulf College Case Study: Technology Enhanced Students' Learning geared with Mobile Devices".* AUEIRC 2017, UAE-Dubai.
US Census Bureau (2011). *An Older and More Diverse Nation by Midcentury.*
Wagner, Tony (2008). *The Global Achievement Gap: Why Even Our Best Schools Don't Teach the New Survival Skills our Children Need – and What We Can do About it.* New York: Basic Books.
Wahba, E. (1998). "*Teaching Pronunciation–Why*?" Language Teaching Forum 36/3: 32. Willems, G. 1987 Communication strategies and their significance in foreign language teaching, System 15: 351–364.

Smart Weather Station

Student Names : Al Zahraa Al Aufi , Reem Al Qamshoui , Hawra al Balushi , Maryam Al Jufaili | Supervisor Dr. Najah Al Mahna, Ali Al-Humairi
German University of Technology Oman, Muscat , Oman

Problem / Question

- The late arrival of the materials .
- Programing of the Xbee transmitter is time consuming.
- One of the sensors burned out and we had to change it .
- We faced difficulties in programing the Grove-Gas Sensor so we couldn't use it.

Hypothesis

- Having a weather station box that includes weather instruments inside it.
- Using smart sensors like(DHT11) instead of using traditional (Thermometer) for measuring temperature.
- Changing the traditional way of checking the result manually by attaching Xbee transmitter that sends data wireless to our device.

Project Overview

The Stevenson shelter was invented in 1864 and it is box placed outdoors. It is designed in a way that no weather conditions such as sun radiation, strong wind or rain can influence its accuracy, while still allowing the air to circulate inside the box so that weather measurements can be made. This shelter has a thermometer to measure the temperature and a hygrometer to measure humidity. This basic shelter has been already developed by adding weather instruments. The aim of our project is to send the readings data from the weather instruments wireless to our devices by using the (Xbee) transmitter and by adding one more sensor (MQ2), which measures the level of CO and smoke.

Variables / Research

- We did a lot of research regarding weather station starting with the sensors , coding of the sensors and ending with making circuits for sensors with the Arduino shield. We found out that many sensors can be useful for our project . The sensors we used are DHT11 (two- in – one sensor), MQ2 . As an improvement for Stevenson shelter we added Xbee transmitter to transfer data to our devices wireless. Having LCD for visualizing the data that we get from sensors was also a good improvement.

Materials

Materials (detailed list)	Quantity (be specific)
Sheets of wood	Six Sheets
Aluminum Stands	Four Stands
White paint	One
LCD	One
Breadboard	One
Arduino	One
Sensors: DHT11(humidity and air temperature sensor) , MQ2(level of CO and smoke). Transmitter : Xbee	One of each sensor One transmitter

Procedure

Step 1	Step 2	Step 3	Step 4
 Propose project and search for further information	 Make Desiccation and agreement	Start manufacturing the body of the project and programing the sensors.	Attach the programming part with the body ten with a computer

Data / Observations

The data we get from this project are :

- DHT 11 : it is humidity – temperature sensor.
  1. Detects temperature in Celsius and Fahrenheit.
  2. Detects humidity in percentage (%).
- MQ2 : detests the level of smoke and CO in the air .
- Xbee : it is a transmitter which sends data wireless from the sensors to our computer device.
- LCD : a screen that shows the data that sensors detect.

Results

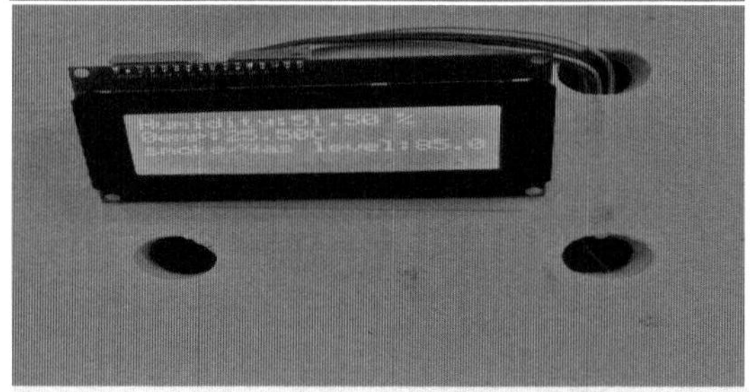

The results we get are shown in the LCD :

- Result 1: shows humidity in percentage.
- Result 2: shows temperature in Celsius and Fahrenheit.
- Result 3: shows CO and smoke level .

Conclusion

In the future we will try to improve the project by:

- Adding more sensors such carbon dioxide sensor
- Use a SIM card to send all of the data from the shelter to our device, for transmission distances of more than 100 m
- Convert all of the readings from numbers to line charts so the decreases and increases in the readings can be seen clearly.
- Change the voltage supply from battery to solar cells

Works Cited

- "Stevenson screen." A Dictionary of Plant Sciences. . Encyclopedia.com. 25 May.2017<http://www.encyclopedia.com>.
- Steadman, R. G. "The Assessment of Sultriness. Part I: A Temperature-Humidity Index Based on Human Physiology and Clothing Science." Journal of Applied Meteorology 18.7 (1979): 861-73.
- Henshall, W. R., and W. P. Snelgar. "A small unaspirated screen for air temperature measurement." New Zealand Journal of Crop and Horticultural Science 17.1 (1989): 103-07. Web.

Graduate Entrepreneurship: Perspectives and Approaches with special reference to HEIs in Oman

Prekerthi Panikar

Caledonian College of Engineering, Muscat, Oman email: panikar@caledonian.edu.om

Entrepreneurship education has received growing attention in recent years due to the perceived benefits it brings to the economy. Many universities have integrated entrepreneurship education as a mandatory course in the university programme. However, researchers have voiced the need to study Graduate Entrepreneurship education differently from the seasoned entrepreneurship courses. This paper aims to shed light on why Graduate Entrepreneurship education should be treated differently and how the approach impacts Graduate Start- ups. The paper takes on a mixed method drawing inference from both empirical study as well as qualitative analysis drawn through Focus Group interviews.

The findings highlight a need for a structured approach to entrepreneurship education for graduates. A three months long course with assessment within a semester of a four year study is probably not likely to get us the results we expect if these startups were to sustain. The researcher suggests the framework for a high impact Graduate Entrepreneurship education in Oman that would yield increased motivation, positive intention for venture start-up and the actual launch.

This research presents some of the challenges educators face in conducting longitudinal studies of graduate startups in Oman and establishes that further longitudinal study is required to assess the survival rates of these startups. Further and ongoing research on mechanisms for ongoing support to these Graduate Startups from the support groups will form a firm base for future studies.

Keywords: Graduate Entrepreneurship, 21st Century skills, Creativity, Innovation, Opportunity recognition, Entrepreneurship as career choice.

References:

Atef, T.M. and Al-Balushi, M., 2015. Entrepreneurship as a means for restructuring employment patterns. Tourism and Hospitality Research, 15(2), pp.73-90.

Bell, R., 2016. Unpacking the link between entrepreneurialism and employability: An assessment of the relationship between entrepreneurial attitudes and likelihood of graduate employment in a professional field. Education+ Training, 58(1), pp.2-17.

Belwal, R., Al Balushi, H. and Belwal, S., 2015. Students' perception of entrepreneurship and enterprise education in Oman. Education+ Training, 57(8/9), pp.924-947.

Burton, M.D., Sørensen, J.B. and Dobrev, S.D., 2016. A careers perspective on entrepreneurship. Entrepreneurship Theory and Practice, 40(2), pp.237-247.

Ennis, C.A., 2015. Between trend and necessity: Top-down entrepreneurship promotion in Oman and Qatar. The Muslim World, 105(1), pp.116-138.

Galloway, L. and Brown, W., 2002. Entrepreneurship education at university: a driver in the creation of high growth firms?. Education+ Training, 44(8/9), pp.398-405.

Matriano, M.T. and Suguku, D., 2015. Entrepreneurship Growth in Oman: Position, Prospects and Growth of Entrepreneural Education. IJAEDU-International E-Journal of Advances in Education, 1(2), pp.127-131.

Nabi, G. and Liñán, F., 2011. Graduate entrepreneurship in the developing world: intentions, education and development. Education+ training, 53(5), pp.325-334.

Nabi, G., Liñán, F., Fayolle, A., Krueger, N. and Walmsley, A., 2017. The impact of entrepreneurship education in higher education: A systematic review and research agenda. Academy of Management Learning & Education, 16(2), pp.277-299.

Smart Irrigating Robot

Nasser Al Harmali, Said Al Badwawi, Budoor Al Bulushi. Supervisor: Dr.Ali Al Humairi, Gutech

Problem / Question

How to save water while irrigating and how to make sure thst your plants irrigated properly.

Hypothesis

The smart AVG irrigate plants when the soil is dry, the AVG has an arm for watering and it has its own tank to save water. When the water is low in the tank, sound as alarm will be activated.

The AVG communicate with the plant by wireless, when it receive a signal from the plant for watering, the robotic car will be activated.

Project Overview

- Now days, due to busy routine life, people forget to water their plants. In this paper, we present a completely autonomous and a cost-effective system for watering indoor potted plants placed on an even surface. The system comprises of a mobile robot and a temperature-humidity sensing module. The system is fully adaptive to any environment and takes into account the watering needs of the plants using the temperature-humidity sensing module. The paper describes the hardware architecture of the fully automated watering system which uses wireless communication to communicate between the mobile robot and the sensing module. This gardening robot is completely portable and is equipped with a Radio Frequency Identification (RFID) module, a microcontroller, an on-board water reservoir and an attached water pump. It is capable of sensing the watering needs of the plants, locating them and finally watering them autonomously without any human intervention. Mobilization of the robot to the potted plant is achieved by using a predefined path. For identification, an RFID tag is attached to each potted plant. The paper also discusses the detailed implementation of the system supported with complete circuitry. The paper concludes with the system performance including the analysis of the water carrying capacity and time requirements to water a set of plants.

Variables / Research

Controlled variables	Independent variable	Dependent variable
• Plants Locations	• Humidity	•Measuring the Humidity in the soil.

Materials

	Part	Quantity	Price	description
1	Servo motors	1	20	Standard Size High Torque Metal Gear Servo
2	DC motors	3	60	12v motors.
3	Wheels	6	150	Moves to X and Y direction.
4	TETRIX parts	1	400	For all robot set
5	Jumpers	50	7	[illegible]
6	DC motor controller	1	5	control X and Y direction motors
7	Infrared tracking module	2	20	Smart car or robot line sensor (including black and white lines)
8	Arduino Uno	2	60	It is place where to create programming language in a simplified version of C/C++.
9	Arduino Mega	1	40	It is place where to create programming language in a simplified version of C/C++
10	Battery	2	50	[illegible]
11	White tape	1	1	For experiment testing.
12	Black tape	1	1	For experiment testing.
13	Xbee module	2	20	S2 module type WIFI communication
14	Water pump	1		Mini Diaphragm Water Pump 2L/min 5 to 12 V For Arduino
15	Water flowmeter	1	3	20mm diameter, <1.75Mpa water pressure and ~30 L/m flow rate range.
16	Humidity sensor with amplifier	1	3	Module dual output mode, digital output, and analog output more accurate. Operating voltage 3.3V-5V
17	Relay	1	5	Switch contactor
18	Step down voltage regulator	2	12	Reduce voltage rate
19	Plastic water tank	2	5	50cm * 30cm box
20	Total cost		862	

Procedure

Step 1

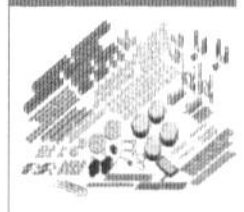

Step 2

Step 3

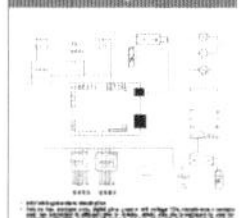

Step 4

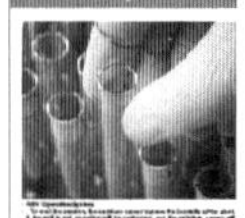

Results

- After completing the design and the gathering of the parts of an AGV, it has been met the objective. All the requirement have been applied to complete and finalize the structure of an AGV, it has been tested and the result was as required, that the AGV to move forward by following a black line, the AGV has been designed to change its direction from X-axis to Y-axis without the front wheel turning, by driving the X-axis motor. The AGV was moving in an accurate motion and stands after each time period to perform its tasks. There were some errors in sensing the black line, and it cause for the AGV to stop or change its direction, and by tuning the range of detection, the sensors were able to detect the line and follow the path without any errors. For the programming, it has been selected as mentioned in previous chapters to use Arduino, the cables connections from the controllers to the Arduino were quite challenging, because of a single mistake can damage any electrical part. It was complex to program the AGV and upload it in Arduino to run the motors with only two sensors, but with the help of Arduino library the program was completed with perfect results. For designing the robotic arm, it has been selected some of Tetrix parts and re-design it, in order to fit the AGV's objectives, the robotic arm were able to move with 5 degree of freedom, and maestro servo controller was used to program the robotic arm, the programming was controlled through the control center in a PC.
- After testing the AGV, the result was successful and the movement was accurate, moreover the speed was low before changing the batteries, because of low voltage and currant, but after replacing the batteries the speed was suitable, and the AGV can be run for a bigger field or for agriculture environment.
- Overall the results were successful and the AGV can accomplish the task. The robotic arm was powerful and precise to pick up the irrigation sprinkler and replace it in a desired position.

Conclusion

- In this paper, we presented a completely autonomous system which helps in watering indoor potted plants that are arranged along a predefined path. The mobile robot is capable of performing three main functions of sensing the watering needs of the plants in need of water, locating them and finally watering them autonomously without any human intervention. The system comprises of the autonomous vehicle, Arduino microcontroller, the RFID reader module, the water pump, the relay, temperature and humidity sensor (HSM-20g), and the Xbee device for wireless communication.
- The AGV has been built to move in both x and y direction by the utilization of five Omni wheels and three DC motors. And the robotic arm is designed to have three degrees of freedom and joint with quarter size pipe to feed water to crops.
- The robot has proven its success when it was put in experiment of simple scenario where the AGV has to follow a black and white line while transporting for crops irrigation from the beginning of the pathway to the end of it.
- For AGV, the sensors that has been used is perfect in detecting the tape and sending signals to Arduino, in order to control the motors, and it has been tested indoor as it is in a farm, but if the AGV is tested in outdoor where the main purpose of this thesis.
- The black tape can be unclear if the AGV was tested outside in a farm, therefore; the sensors cannot detect the black line and the AGV will directly stop, unless the black tape is cleaned. In agriculture environment there are unequaled level of the field, and heavy sand that can stop the AGV with two wheel drive, therefore to improve the motion of the vehicle, few changes need to be done in order to keep the motion of the AGV more relevant.

A STUDY ON CAUSAL NEXUS BETWEEN GDP AND FDI - THE CASE OF OMAN

Dr.M.Tamilselvan
Business Studies Department.Ibri College of Technology
Ibri, Sultanate of Oman- tamilindias@yahoo.co.in

Mr.A.Jayakumar
Business Studies Department.Ibri College of Technology
Ibri, Sultanate of Oman.

Abstract-The review of literature evidently proves that the international funds are the significant factor for the development (GDP) in many countries. It supports the developing countries in capital formation, technology up gradation, gaining adequate foreign exchange and knowledge transfer to enhance the growth. In this direction the present study tests the causality of international investment on GDP in Oman. The Johansen integration, Restricted Var Model and wald statistic were employed to study the subsistence of independent and joint connection in short and long run between 1998-2016.The empirical results derived from VECM and wald test indicate the nonexistence of joint and independent causality in short run and long run.

Key Words: FDI led Growth Strategy –Causality – Cointegration -Vector Error Correction Model (VECM)

Introduction

Oman is confronting challenges due to oil price turmoil. The budget deficit was $6.5 billion in 2015 around 11% of GDP. Hence, it focuses on reducing the oil sector contribution to GDP from 46% to 9% by 2020 through diversification, industrialization and privatization. There are few challenges to be addressed such as creating job to the young workforce, maintaining social welfare spending and sustaining growth from non oil sector contribution.[1]It has been evidently proved in many countries that theinternational capitalhas been instrumental for gross domestic production. It helps to improve the capital formation and technology incorporation to streamline the firms. It is actively bridging the gaps between savings and investment, foreign exchange and tax-revenue gap. Besides FDI initiates the knowledge and asset transfer to improve the productivity of the recipient nations. (Pradhan,2009)[2]. In Oman, the FDI has been channeled about $39128 million in various sectors such as Coal, Oil and Natural Gas, Metals, Chemicals, Real Estate and Hotels & Tourism (The top five sectors) between 2003-2015)[3]. In this direction this paper is investigating the role of FDI on GDP in Oman.

The review of literature is guiding to estimate the causality in two approaches such as production function and time series approaches. The present study is opting the second method to assess the underlying relationship of international capital on GDP in Oman. There are many previous studies brought out the mixed results such as unidirectional, bidirectional and nonexistence of interconnection. In spite of several works achieved in the area, only very few studies have been conducted in GCC region especially in Oman and the study included recent data to test the causation. The remaining part of the study is structured as follows: section 2 describes literature review; section 3 provides methodology 4 results and discussion; and section 5 provides conclusion

Literature Review

FatehHabibi and Mohammad Sharif Karimi (2017)examinedthe influence of international capital oneconomic growth in GCC countries for

[1]https://theodora.com/wfbcurrent/oman/oman_economy.html

[2]Rudra Prakash Pradhan (2009)The FDI- Led- Growth Hypothesis in ASEAN- 5 Countries: Evidence from Cointegrated Panel Analysis, International Journal of Business and Management. Vol. 4, No. 12

[3] : FDI Intelligence from The Financial Times

the period 1980-2014 using ARDL approaches and revealed that there is extensive steady-state affiliation between international capital and GDP in Iran and for individual country of GCC. The Granger-causality test inferred that a bidirectional causality from international capital to real GDP growth in Qatar, Saudi Arabia and UAE; unidirectional causalities from FDI to real GDP growth rate in Iran and Bahrain and no causation between international capital and real GDP growth rate in Kuwait and Oman.

Seng Sothan (2017) studied the underlying association between growth and international capital between 1980–2014, adopting using Granger causality model. The outcomes revealed a robust contributing impact of international capital on Cambodia's economic growth.

Gabriela Markova (2016) found unidirectional causality from exports to GDP in Greece and Italy, one-directional causality from GDP to exports in Denmark, Germany, Korea, New Zealand, the Netherlands, Canada and Norway, bidirectional causality between exports and growth in Japan, Sweden and Turkey, while in the cases of Australia, Austria, France, Iceland, Ireland, Mexico, Spain, Switzerland and the United States there is no statistical evidence for any causal relationship.

Adeleye and Adeteye (2015)examined the effect of global trade on gross domestic production in Nigeria between 1988-2012 employing co-integration and error correction model and confirmed the presence of undeviatingassociationamong export and growth in Nigeria.

FurrukhBashir (2015)investigated the effect of Export on gross domestic production in Pakistan between 1972-2012 by using stationarity test, Cointegration, restricted var model and Granger causality tests. The results exposed that there is a solid optimistic extensive and diminutive correlation between exports and economic growth in Pakistan.

Muhammad Tahir (2015)attempted to review the affiliation between export and growth in Sri Lanka for the period 1981 to 2012 using co integration and VAR model and the results found that the non-existence of association between the export and growth in SriLanka.

NemanjaLojanica (2015) studied the causality between income and expenditure in the Republic of Serbia employing autoregressive distributed lag model (ARDL). The observedoutcomesdiscovered that there is a co integration between income and expenditure and unidirectional causality from expenditure towards income.

Karunanithy and Ramachandran (2014) examined the causal relation between Bombay Stock Exchange sensitive index and gold price adopting Granger Causality test. The result observed the occurrence of unidirectional causal relationship from index to gold price.

Deepika Kumari and Neena(2014) Studied the underlying association among exports and economic development by engaging Johansen co integration and Granger causality models for the period of 1980 -2012. The results determined that there is no long run association between exports and GDP.

SeyedMohammadrezaHosseini (2014) studied the association among Export and Growth in India for the period of 1960 to 2010 using the co-integration and Granger causality methods. . The results exhibited the presence of unidirectional interconnection among export and economic development for India. .

MohammedIssa(2014) studied the correlation involving energy consumption and real economic development in seventeen Arab nations from 1980-2011 by means of Auto Regressive Distributed Lag (ARDL): The findings confirmed that the absence of interconnection from economic growth to energy consumption.

Ibrahim Alomar(2013)studied the association between domestic reserves and economic growth among Gulf Cooperation Council nations between 1980-2010 employing Granger causality test. The findings concluded that the developmentspeedroots savings in 4 countries. The contradictory results happened inOman and bi directional causationwas found in Bahrain.

Kamal and Jobaer(2013)studied the association between international trade and growth in Bangladesh from 1981-2013 adopting OLS method. The investigation revealed the reasonable correlation between the variables. Export is contributing positive effect on growth but import's influence is indifferent.

MahmoudAbolpourMofrad (2012)studiedthe long and short run correlation between GDP, export and investment in Iranfrom 1991-2008employing restricted var model.The inferencesshown that there occurs a constructive and substantial long runaffiliation between investment and export with growth and found negative relationship between investment and export.

Mohammed Ameen Fadhil, and WalaaIsmeal (2012)investigated the underlyinginteractions between international capital and growth in Qatar during 1990-2010 using VAR and IRF model. The results confirmed the bi-directional interconnection and extended interactions between international capital and growth in Qatar.

Mohsen Mehrara(2011)studied the association of export (except oil) and development adopting panel cointegration investigation for 73 emergent nations from 1970-2007. The outcomesestablishedthat there is bidirectional extended interconnection between export and growth.

BushraRiaz (2010)explored the contributoryaffiliationbetween exports and growth by employing the time series model from 1980-2009 for 13 developing nations. The analysis observed the presence of reciprocated extensive association among exports and growth in maximum countries.

Khaled (2010)tested the association between the exports and growth in Libya during 1980-2007 adopting cointegration; Granger causality tests. The result specified that the export strategy provided to the progress within Libya.

Yuhong Li, Zhongwen Chen, Changjian San (2010)measured the underlying consequence among global trade and development within China from 1981-2008 engaging co integration and restricted var model. The result recommended the presence long and short run causation between gross domestic production and international trade.

TomislavGelo (2009) investigated the association among economic development and energy utilization in Croatia during 1953-2005adopting Granger causality test. It showed the incidence of mutual causality between energy consumptions and gross domestic production.

Mahmoud Al-Iriani(2007) tested the linkage between international venture and development in the six nationsincluding the Gulf Cooperation Council (GCC) during 1970-2004 using Granger causality tests.The findings confirmed that there is bidirectionalinterconnection between international investment and growth among GCC nations.

Choong Chee Keong(2005) Studied the correlation between growth and export in Malaysia during 1960-2001 using bound test .The findings presented that the exports impacts growth in Malaysia.

Ugo Fasano and Qing Wang (2002) examined the connection between total state expenditure and income in GCC states during 1975-2000adopting co integration and restricted varmodeling framework. The findings confirmed that the government spending follows oil revenue.

Out of 23 reviews briefly described above, only four studied have been conducted to test the causality between FDI and GDP from which only three studies were performed including GCC nations. Besides most of the studies are related to GDP and Export, Import and so on.Mohammed Ameen Fadhil, and WalaaIsmeal (2012) and Mahmoud Al-Iriani(2007) studied using FDI and GDP and focused on Qatar and GCC in general. FatehHabibi and Mohammad Sharif Karimi (2017) was the only study tested the causality of FDI on GDP on GCC nations and revealed the nonsubsist of causal relationship between FDI and GDP in Oman using ADL bound approach. Therefore, the entire review observed that no exclusive study was conducted to test the causality of FDI on GDP in the recent past using VECM. Hence, the current study is motivated toexamine the causal impact of international investment on Gross Domestic Producxtion using VECM focusing on Omanexclusively.

Methodology

The paper used secondary facts of Gross Domestic Production and Foreign Direct investment for 19 years from 1998 to 2016 in Oman retrieved from https://tradingeconomics.com/.Thepaper adopted Johansenco integration, restricted var model and Wald statistics to find the independent and joint causality of FDI on GDP in both the short run and long run. Besides, the stationarity of the sample data was tested using appropriate investigation. The data analysis has been done through E-view -9.5 econometrics software in this research.

Johansen – Juselius Co integration

$$\Delta\gamma_t = \sum_{i=1}^{p=1} \Gamma_i \Delta\gamma_{t-i} + \Pi\gamma_{t-i} + \varepsilon_t \text{ ------}(1.1)$$

Where γ_t is the 2x1 vector (FDI, GDP) respectively, Δ is a symbol of difference operator, ε_t is a 2x1 vector of residuals. The VECM model has information about the short and long run adjustment to changes in γ_t via the estimated parameters Γ_i and Π respectively. Here, the expression $\Pi\gamma_{t-1}$ is the error correction term and Π can be factored in to two separate matrices α and β, such as $\Gamma = \alpha\beta$, where β denotes the vector of co integrating parameters while α is the vector of error correction coefficients measuring the speed of convergence to the long run steady state.

Vector Error Correction Model (VECM)

$$\Delta GDP_t = \sum_{i=1}^{p=1} \beta_i \Delta GDP_{t=i} + \sum_{i=1}^{p=1} \alpha_i \Delta FDI_{t-i} + Z1 * EC1_{t-1} +$$

Where β_i and α_i are the short run coefficients, EC1 is error correction terms, ε_{1t} is the residuals in the equation 1.2.The $EC1_{t-1}$ is the lagged value of the residuals derived from the co integrating regression of GDP on FDI (Equation 1.2).

Results and Discussion

The calculated p-values of FDI in both ADF and Phillip-Perron tests are 0.05. Therefore, the null hypothesis is not accepted and ensured that the sequence is stationary at both level and first variation. In contrast, the calculated p-values of GDP are more than 0.5 at level in both the tests and less than 0.05 at first difference in both the tests. Hence the null hypotheses are not rejected at level and rejected at first difference. It ensures that the sample GDP facts is not immobile at level and becomes stationary at first difference. After performing unit root test, the existence of co-integration is tested using Johansen Co-integration test. The calculated p-values in both Trace statistic and Max-Eigen statistic are less than 0.05 at none and more than 0.05 in at most one. Hence, it is found that the co -integration exists between the variables GDP and FDI. Therefore, the restricted VAR model (VECM- Vector Error Correction Model) is adopted to test the long and short run joint and independent causality of FDI on GDP.

The calculated p-values of VECM L(1) 0.9338, S(2) 0.6145, S(3) 0.8171,S(4) 0.2403, S(5) 0.5506, S(6) 0.0560, S(7) 0.2676 and S(8) 0.4484. All the values are more than 0.05. In the mentioned values L(1) indicates the long run association between GDP and FDI. Since the value is more than 0.05 it is confirmed that there is no long run relationship between GDP and FDI. S(2), S(3) and S(4) are the lagged GDP variables which indicate the short run independent relationship. It shows that the causal relationship is not found between GDP lagged GDP variables. Similarly the calculated p-values of lagged FDI S(5), S(6) and S(7)] are showing the short run causality of FDI on GDP. The p-values are more than 0.05 and not supporting the short run independent causality. Finally, the wald test is used to discover the presence of short run joint association between GDP and lagged FDI variables. Since the calculated p-value of wald test is more than 0.05 it is confirmed that the lagged variables S(5), S(6) and S(7) are not jointly having causal relationship on GDP.

Though the non-existence of short run and long run joint and independent association among GDP and FDI, the calculated R-Square is more than 65% and p-vale of F-statistic is significantly validate the model. So, it is required to perform the residuals investigative tests such as auto correlation, heteroskedasticity and normality of residuals. The p-values of Breusch – Godfrey Serial Correlation LM Test, BreuschPagan – Godfrey Heteroskedasticityand Normality Distribution of Residuals -Jarque-Bera are 0.0144, 0.2125 and 0.8109. Among three values the first value is less than 0.05 and the null hypothesis that the residuals are not serially correlated is rejected which means that there is a autocorrelation in the residuals. On the other hand the other two values are more than 0.05 and accept the null hypotheses that the residuals are not suffering from heteroskedasticity and the residuals are normally distributed which emphasize that there no heteroskedasticity in the residuals and it is normally distributed.

Summary and Conclusion

The study attempts to inspect the causal consequence of international investment on GDP in Oman for the period of 1998 -2016 using restricted var method and coefficient diagnostic test wald statistics. The results of the VECM disclose that the foreign direct investment is neither causingGDP in short run nor in the long run. The short run joint relationship not identified in wald test. Therefore, it is concluded that the foreign direct investment is not making any impact on the GDP in Oman.It supports the findings of FatehHabibi and Mohammad Sharif Karimi (2017). That means, the FDI has not been adequately propelled in various non oil sector in order to boost the growth in Oman. Eventually, the study suggest the policy makers to be straight forward in FDI policy to attract as much as possible to achieve diversification, industrialization and privatization to reduce the oil sector growth contribution from 46% to 9% by 2020.

References

Adeleye J. Adeteye O. &Adewuyi M. (2015) Impact of International Trade on Economic Growth in Nigeria (1988-2012) International Journal of Financial Research Vol. 6, No. 3; 2015

BushraRiaz (2010) Comparative Advantage, Exports and Economic Growth, Master Program Thesis, Department of Economics, SödertörnsHögskola

Choong Chee Keong(2005) Export-Led Growth Hypothesis In Malaysia: An Investigation Using Bounds Test, Sunway Academic Journal 2, 13–22 (2005)

DeepikaKumari(2014) Export-Led Growth in India: Cointegration and Causality Analysis**,** Journal of Economics and Development Studies June 2014, Vol. 2, No. 2, pp. 297-310

FatehHabibi and Mohammad Sharif Karimi (2017) Foreign Direct Investment and Economic Growth: Evidence from Iran and GCC, Iran. Econ. Rev. Vol. 21, No. 3, 2017. pp. 601-620

Furrukh Bashir(2015) Exports-Led Growth Hypothesis: The Econometric Evidence From Pakistan, Canadian Social ScienceISSN 1923-6697,Vol. 11, No. 7, 2015, pp. 86-95

Gabriela Markova (2016) Granger Causality Between Exports and Growth in OECD Countries: A Panel Data Approach, Bachelor Thesis in Economics, International Business School, Jonkoping University.

Ibrahim Alomar(2013) Economic Growth and Savings in GCC:A Cointegration and Causal Relationship Analysis, International Journal of Humanities and Social Science Vol. 3 No. 9; May 2013

Karunanithyand Ramachandran (2014) Causal Relationship between Stock Price and Gold Price in India: A Granger Causality Test Approach, International Journal of Research in Management, Science & Technology (E-ISSN: 2321-3264) Vol. 2, No. 2, August 2014

Khaled (2010) The Relationship between Export and Economic Growth in Libya Arab Jamahiriya Theoretical and Applied Economics Volume XVII (2010), No. 1(542), pp. 69-76

Muhammad Tahir(2015) An Analysis of Export Led Growth Hypothesis: Cointegration and Causality Evidence from Sri Lanka, Advances in Economics and Business 3(2): 62-69, 2015

Mohammed IssaShahateet(2014) Modeling Economic Growth and Energy Consumption in Arab Countries: Cointegration and Causality Analysis,International Journal of Energy Economics and Policy Vol. 4, No. 3, 2014, pp.349-359 ISSN: 2146-4553

Mohsen Mehrara(2011) Granger Causality Relationship between Export Growth and GDP Growth in Developing Countries: Panel Cointegration Approach, International Journal of Humanities and Social Science Vol. 1 No. 16;

Mahmoud AbolpourMofrad (2012) The relationships between GDP, export and investment: case study Iran, Business Intelligence Journal.

Mahmoud Al-Iriani (2007) Foreign Direct Investment and Economic Growth in the GCC Countries: A Causality Investigation Using Heterogeneous Panel Analysis, Topics in Middle Eastern and North African Economies, electronic journal, Volume 9, Middle East Economic Association and Loyola University Chicago, September, 2007, http://www.luc.edu/orgs/meea/

MOHOHAMMED AMEEN FADHIL, LIU YAO, WALAAISMEAL(2012) CAUSALITY RELATION BETWEEN FDI INFLOWS AND ECONOMIC GROWTH IN QATAR,INTERNATIONAL JOURNAL OF BUSINESS AND MANAGEMENT, ISSN 1833-3850 (PRINT) ISSN 1833-8119 (ONLINE)

NemanjaLojanica (2015) Government Expenditure and Government Revenue – The Causality on the Example of the Republic of Serbia, Management International Conference,Portoroz, Slovenea.

Seng Sothan (2017)Causality between foreign direct investment and economic growth for Cambodia,https://doi.org/10.1080/23322039.2016.1277860

SK Kamal,Md. AnamulHoqueand(2013)Effects Of Export And Import On GDP Of Bangladesh An Empirical Analysis,The International Journal Of Management- Vol 2 Issue 3 (July, 2013)ISSN 2277-5846

SeyedMohammadrezaHosseini(2014) An Empirical Study of Export and Economic Growth in India since 1960: A Co integration Analysis, Iran. Econ. Rev. Vol.18, No. 1, 2014.

TomislavGelo (2009) Causality between economic growth and energy consumption in Croatia Zb. rad. Ekon. fak. Rij. • 2009 • vol. 27 • sv. 2 • 327-348

Yuhong Li, Zhongwen Chen, andChangjian San (2010) Research on the Relationship between Foreign Trade andthe GDP Growth of East China—EmpiricalAnalysis Based on causality, Modern Economy, 2010, 1, 118-124.

Shri Harun R Khan, Deputy Governor, Reserve Bank of India, Impact on Indian Economy, PHD Chamber of Commerce. 12.03.2016.

Ugo Fasano and Qing Wang(2002) Testing the Relationship Between Government spending and revenue: Evidence from GCC Countries.IMF working paper, JEL classification Numbers:E62,H50,N15.

Incorporation of Case Approach in Teaching Entrepreneurship Course at Higher Education Institutions

Raihan Taqui Syed [1*], Sariya Nasser Al Ismaili [1], Abeer Al Balushi [1]

[1] Modern College of Business & Science, Muscat, Oman

(Affiliated to University of Missouri, St. Louis, USA)

*Corresponding Author: Email address – raihan@mcbs.edu.om

Entrepreneurship is a well-established, dynamic medium which has resulted in nations developing and prospering. There are umpteen cases of innovation and creativity driving growth of organizations and subsequently creating wealth in the countries. Emphasis is being given to Small & Medium Enterprises (SMEs) as these are regarded to be the core of creativity & innovation, and platform for further economic development and diversification.

The significance of SMEs & Entrepreneurship in diversifying the Omani economy was realized and under the directives of His Majesty Sultan Qaboos, where the entrepreneurship course was created. This course was made mandatory to be taught at undergraduate level across all Higher Education Institutions (HEIs) in Oman. The guidelines provided by Ministry of Higher Education are comprehensive and include curriculum, course offering template and extra-curricular activities.

Entrepreneurship as a subject of study at various HEIs across the world is positioned in different stages. In Oman, it is still in the nascent stage as the course was introduced only in 2014 across all HEIs. There are many tools for instructors to utilize as a way to minimize the many obstacles faced by teaching such course in a country where entrepreneurship is an extremely young and growing culture. One of the major ways of maneuvering through the obstacles faced whilst teaching entrepreneurship to HEI students is through the Case Approach. The paper analyzes the issues regarding the Case Approach and details how to eliminate these challenges & what instructors can do to fully benefit from such approach towards instilling an entrepreneurial culture across today's generation in Oman.

Key words: *Entrepreneurship, Higher Education Institution, Case Approach*

Revisiting the teaching materials of an introductory programming course

Sohail Iqbal Malik, Mohanaad Shakir, Abdalla Eldow, Mohammed Waseem Ashfaque
Department of Information Technology
Buraimi University College
Al-Buraimi, Oman
{Sohail,Mohanaad,Abdalla,Ashfaque}@buc.edu.om

Abstract—**Computer programming is considered as a difficult area of study for novices. One of the reasons is the main focus of the curriculum presented in an introductory programming (IP) course which emphasizes more on the programming knowledge (syntax and semantic) of the programming language. This study introduced a new teaching curriculum in the IP course which focuses on different skills required by the novices. We compared the IP course materials based on the traditional and new approaches against five categories. The result shows that the new approach encourages both the programming knowledge and problem solving strategies, and promotes deep learning. Furthermore, it discourages programming shortcut (Problem statement → Code), and support three-step approach (Problem statement → Solution Plans → Code) in solving a problem statement.**

Keywords—Introductory programming; Novice programmers; Teaching materials; Problem solving strategies;

I. INTRODUCTION

The introductory programming course is considered as one of the difficult courses for novice programmers [1]. They have to focus on different skills, such as program comprehension, problem solving strategies, program design etc., at the same time. On the other side, the traditional teaching approach used in the IP course focuses more on the syntax of the programming language [2]. This study evaluated the teaching materials of the IP course based on the traditional approach. A new teaching approach based on the ADRI (Approach, Deployment, Result, Improvement) model was introduced in the IP course to address the shortcomings emerged from the traditional approach. To determine the impact of the new teaching approach, the teaching materials based on the traditional and new approaches were analyzed by the criteria proposed by [3].

This paper is organized into a number of sections. It starts with a literature review followed by a brief introduction to the new teaching approach. Research methodology and Results of the study were presented. The paper concludes with the research outcomes of the study.

II. LITERATURE REVIEW

Various studies have been conducted to investigate the different teaching approaches used in IP courses. In this section, we explore the literature and discuss different teaching approaches.

The teaching method is an approach by which educators deliver teaching materials to the students. A teaching strategy is one of the crucial factors that help to motivate students and ultimately play an important role in mastering the course content [4]. Students' achievement of the learning outcomes is affected by their learning style and motivation. Some students prefer a visual learning style, others prefer auditory or kinesthetic styles [5] and therefore it is important for teachers to incorporate different approaches to accommodate different learning styles in their teaching process.

Reference [6] interviewed instructors of an IP course to learn about the pedagogical and motivational strategies they used in the course for the students. They interviewed 18 instructors and concluded that 'student-teacher relationship, teachers' proximity with the students, class competitions, challenges and strategies in which the students are engaged actively and continuous assessment' (p. 7) are considerable motivational strategies. Moreover, most teachers' do not consider this course different from other courses and use the same traditional pedagogical strategies in it. Some teachers' emphasized that a good choice of exercises and teaching materials also play an important role in a successful teaching and learning process.

Reference [7] discussed that if instructors omit any one of these models ('data structure, program design, problem domain and data representation') from the teaching process of the programming then as a consequence 'the students will make up their own models of dubious quality' (p. 21).

Reference [8] introduced Assurance of Learning (AOL) process in an introductory programming (IP) course to align program learning goals with the IP curriculum. The five stages of the AOL process helped them to identify shortcomings in the course and provided mechanism to address these drawbacks. They collected faculty feedback regarding the AOL process through a web based survey. The overwhelming majority of the participants agreed that the AOL affected their teaching process in a positive way.

Mobile device supported learning provides additional support for accessing introductory programming teaching materials to students [9]. It provides teaching materials anywhere, anytime and eliminates space and time constraints for learners. It also allows interaction between teachers and learners without physical contact [10]. The work of [11] suggested that the retention of specific subject knowledge in first year university students for a mobile learning group was better than for those who were not supported in this way. Surveys conducted with novice programmers concluded that the overwhelming majority of participants agreed that mobile device supported learning provides additional support to access and understand teaching materials. The teaching materials can be saved in mobile memory which can be accessed at the learners' convenience. Furthermore, the students can access recorded class lectures along with other teaching contents. It can help them to catch-up with missed classes [9]. Moreover, it helps to promote information communication technology as a teaching tool in the teaching and learning process [12][13].

The ACM-IEEE Joint Task Force on Computing Curricula [14] identifies 18 knowledge areas for computing undergraduate programs. Programming knowledge (syntax and semantics) and problem solving strategies are given equal attention in these knowledge areas. The IP curriculum should promote both skills so that students will be well prepared for their career path and labor market. Reference [15] reported that curriculum organization and teaching methods were the two main factors for high failure rates in IP courses. Reference [16] analyzed forty programming books and concluded that most of the books emphasized the syntax of programming languages. Some authors provide a large number of examples but give less attention to problem solving skills. In some cases, authors discuss problem solving skills in early chapters but these skills are not integrated in the remaining chapters.

Table 1 lists the factors and current practices used in teaching and learning process of an IP course. In this study, these factors were addressed by introducing a new teaching approach instead of the traditional approach to enhance the learning process.

The traditional approach emphasizes more on programming knowledge (syntax and semantics), promotes surface learning and programming shortcut (Problem Statement→Codes). Most of the programming editors used in the introductory programming also promote programming shortcut.

Table 1. Factors and practices in traditional approach

Factors	Practices in Traditional Approach
Learning to program	More emphasis is given on syntax and semantics of programming languages
Teaching strategy/model	Traditional approach spends more time on teaching syntax compared to problem solving strategies. Hence does not achieve students' learning outcomes and high failure and dropout rates are reported
Programming books	Most textbooks discuss problem solving strategies in an initial chapter only and syntax and semantics are emphasized in rest of the chapters
Learning style	Traditional approach promotes surface learning instead of deep learning because problem solving strategies are not emphasized thoroughly in all topics
Lecture notes	Examples discussed in the lectures promote programming shortcut (Problem Statement→Codes)
Lab exercises	Problems given in the exercises promote programming shortcut (Problem Statement→Codes)
Software development tools	Most of the editors used in the introductory programming promote programming shortcut (Problem Statement→Codes)
Programming language practice	Most students struggle with completing programming exercises because they spend most of their time in debugging their programs

III. NEW TEACING APPRAOCH

The new teaching approach based on the four stages of the ADRI model was introduced in the IP course. The ADRI approach focuses on different skills (problem solving strategies, syntax, semantics and problem analysis) in the IP course. All the programming examples and problems were prepared based on the four stages of the ADRI approach as shown in table 2. The first part of the ADRI based example provides a problem statement. The second part (Approach) covers problem solving skills such as pseudo code and flow chart. The third part (Deployment) focuses on the syntax and semantics of the programming language. The fourth part (Result) covers input, output, process used to solve the problem statement, and basic syntax and semantic errors in the programming language. The fifth part (Improvement) focuses on new programming constructs, syntax and semantics of the programming language by changing the requirements of the problem statement given in the first part.

Table 2. ADRI based programming example

<table>
<tr><td colspan="2">Design an algorithm that will read name, balance and rate then calculate interest and print name and interest.</td></tr>
<tr><td>Step 1: Approach – Problem-solving strategies
Solve the given problem statement by pseudo code technique</td><td>Solve the given problem statement by flow chart technique</td></tr>
<tr><td colspan="2">Step 2: Deployment – Programming knowledge
Convert the proposed solution in step 1 into a computer program by using the programming language</td></tr>
<tr><td>Step 3: Results
Input:
NAME : Naushad
BALANCE: 20
RATE : 0.25
Process:
INTEREST = 20 x 0.25
Output:
Name: Naushad
Interest: 5</td><td>Syntax Error: Missing Semicolon
Example:
4 int feet;
5 int inches
6 double bigNum;
Reason: Semicolon is expected on line 5</td></tr>
<tr><td colspan="2">Step 4: Improvement
Update the above algorithm to calculate RATE if INTEREST and BALANCE is entered by the user.</td></tr>
</table>

An ADRI editor was prepared to support the four steps of the ADRI approach as shown in figure 1. The editor provides separate interface for each step of the ADRI approach. The editor was developed in Java program. It is compatible with most of the operating system and easy to use. All the programming questions are embedded in it. The user can choose question by topic.

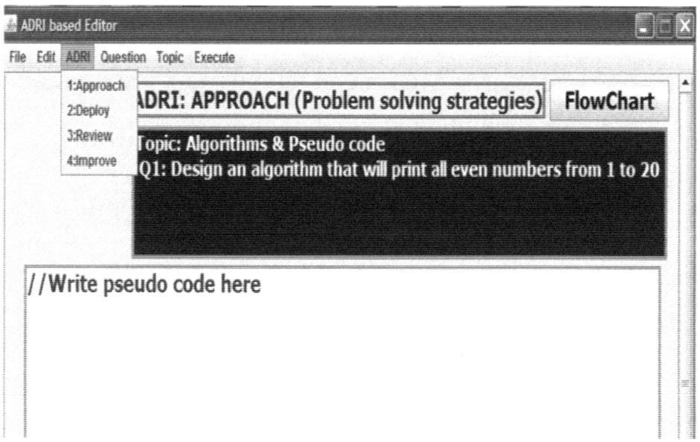

Figure 1. ADRI based Editor [17]

IV. RESEARCH QUESTION

It is evident from the literature that the teaching materials based on the traditional approach do not provide all the required skills to novices. This study purposes the new teaching materials based on the four stages of the ADRI approach in the IP course. Moreover, the study purposes the following research question to determine the impact on the teaching materials based on the ADRI approach compared to the traditional approach.

RQ1. What is the impact of the ADRI approach compared to the traditional approach on the teaching materials of the introductory programming course?

V. RESEARCH METHODOLOGY

The research question was probed by comparing the IP course materials based on the traditional and ADRI approaches against five categories such as *Teaching topics for lecture sessions, Practical topics for lab sessions, Types of programming examples and problems, Presentation style of example and problems,* and *Four problem solving steps in examples and problems*.

VI. RESULT

This section describes the research question. We compare the teaching materials based on the traditional and ADRI approaches to determine the impact of the ADRI approach on the course.

Table 3 below shows the comparison between the traditional and ADRI based teaching materials of the course. The teaching materials are compared in five different categories including: '*Teaching topics for lecture sessions', 'Practical topics for lab sessions', 'Types of programming examples and problems', 'Presentation style of example and problems', and 'Four problem solving steps in examples and problems'*. The comparison of the traditional and ADRI approaches in the first two categories (*'Teaching topics for lecture sessions', 'Practical topics for lab sessions'*) are discussed in terms of the emphasis on the programming skills and problem solving strategies in each category of teaching materials. The remaining three categories (*Types of programming examples and problems, Presentation style of example and problems,* and *Four problem solving steps in examples and problems*) were analyzed on the criteria provided by [3].

Reference [3] presented a content analysis of computer programming examples or problems of sixteen textbooks taught in high school in Taiwan. They categorized the programming problems into five different types such as math problem (finding the even integers), syntax-oriented problems (demonstrating the different programming constructs), real-life problems (how to convert Fahrenheit and Celsius temperatures, calculation of bank loans), graphics problems, and others. Graphic problems are not taught in the IP course and so were not considered in our analysis.

Reference [3] discussed a qualitative analysis of three different presentation styles used for the programming examples in text books. A brief explanation of these presentation styles as follows:

Problem Statement → Pseudo-code or Flowchart: The problem statement is pursued by pseudo-code or flowchart

Problem Statement → Codes: The problem statement is pursued by code with a brief explanation.

Problem Statement → Solution Plan → Coding: The problem statement is pursued by solution plan (problem analysis, input, output, process, program design, variables etc.) and then complete code is given with a brief explanation.

Reference [3] demonstrated problem solving as a 4-step process as follows:

The problem analysis step: The problem statement is described so that programmers can understand it. There is no particular method used for this step.

The solution planning step: It is demonstrated by a textual description such as algorithm design, variables, main algorithm etc.

The coding step: A complete code is given with a brief explanation or comments. A program is also provided with a sample run.

The testing/debugging step: It shows to students how to debug syntax, semantic and run-time errors by presenting programs with planted errors in it.

Table 3. Comparison of the traditional and ADRI based teaching materials

<table>
<tr><th>No</th><th>Area of concern</th><th colspan="2">Traditional approach</th><th colspan="2">ADRI approach</th></tr>
<tr><td rowspan="3">1.</td><td rowspan="3">Teaching topics for lecture sessions</td><td colspan="2">More emphasis on syntax and semantics compared to problem solving strategies</td><td colspan="2">Pay equal attention on syntax & semantics and problem solving strategies</td></tr>
<tr><td>Syntax & Semantics</td><td>Problem Solving Strategies</td><td>Syntax & Semantics</td><td>Problem Solving Strategies</td></tr>
<tr><td>86.6%</td><td>13.4%</td><td>100%</td><td>100%</td></tr>
<tr><td rowspan="3">2.</td><td rowspan="3">Practical topics for lab sessions</td><td colspan="2">More emphasis on syntax and semantics compared to problem solving strategies</td><td colspan="2">Pay equal attention on syntax & semantics and problem solving strategies</td></tr>
<tr><td>Syntax & Semantics</td><td>Problem Solving Strategies</td><td>Syntax & Semantics</td><td>Problem Solving Strategies</td></tr>
<tr><td>86.6%</td><td>13.4%</td><td>100%</td><td>100%</td></tr>
<tr><td>3.</td><td>Types of programming examples and problems</td><td colspan="2">Syntax-oriented examples and problems are dominant in reading materials (50%) and lecture notes (49%)</td><td colspan="2">Math (62.7%) and daily-life (25.6%) examples and problems are dominant</td></tr>
<tr><td>4.</td><td>Presentation style of example and problems</td><td colspan="2">Problem Statement→Codes presentation style is dominant in lecture notes (77%) and reading materials (70%)</td><td colspan="2">Problem Statement→Solution Plans→Codes presentation style is dominant in lectures (83.3%) and labs (84%)</td></tr>
<tr><td>5.</td><td>Four problem solving steps in examples and problems</td><td colspan="2">Coding step is dominant in lecture notes (74%) and reading materials (61%)</td><td colspan="2">All the four problem solving steps are addressed</td></tr>
</table>

It is clear from table 3 that in '*Teaching topics for lecture sessions*' category, the ADRI approach pays equal attention to syntax & semantics and problem solving skills compared to the traditional approach which only emphasizes the syntax & semantics. The same trend is depicted in '*Practical topics for lab sessions*' category. For '*Types of programming examples and problems*' category, syntax-oriented examples are more dominant in the traditional approach compared to math and daily-life which are dominant in the ADRI approach. Problem Statement → Codes style is dominated in the traditional approach. Under '*Presentation style of example and problems*' category compared to the ADRI approach, Problem Statement → Solution Plans → Codes style is dominated. Lastly in '*Four problem solving steps in examples and problems*' category, coding is dominated in the traditional approach compared to the ADRI approach where all the four problem solving steps are dominated. Although the change in emphasis from coding dominating to a more equal presentation of problem solving and coding was the aim of the redesign and modification of the teaching materials, this comparison demonstrates that the goal was successfully achieved.

VII. Discussion

The IP teaching materials based on the traditional and ADRI approaches were analyzed. The comparison shows that the ADRI approach enhanced the students' knowledge and confidence in the programming domain. They realized and grasped the programming concepts in a better context which helps them in developing their programming skills. The multi-stage approach to develop a solution helped students in developing their flow of logic in a structured way and to achieve their objectives in the course. This finding is consistent with [18].

In the ADRI based teaching materials, math and daily-life programming examples and problems are dominant which helped in developing the students' interest in the course. The students already had a good knowledge of math's from their school and current courses so it was comparatively easy for them to solve the problem statement based on it. Moreover, it was easy for the students to understand the requirements of the problem statements based on daily-life scenarios. Syntax-oriented type of programming examples and problems are dominant in the traditional approach which gives good understanding of the programming knowledge to the students but it does not promote problem solving skills.

The traditional approach style (Problem Statement → Codes) promotes programming shortcuts. However, for novices it is very important to practice program design as suggested in some research studies [7][14][16][19]. The ADRI approach discourages programming shortcut and promotes the three-step approach (Problem Statement → Solution Plans → Coding) which develops program design, language features and program comprehension skills. The purpose of the whole is to promote deep learning of the programming domain to the novices. The three-step approach offered by ADRI is consistent with that suggested by [20]-[23].

The step 3 (Result) of the ADRI approach covers input, output, process used to solve the problem statement, and basic syntax and semantic errors in the programming language. This process helps novices in understanding the given problem statement in a better way and promotes deep learning. Common programming errors are presented to novices which help them in writing the code of the program with less errors and saving their debugging time. Furthermore, novices can focus more on the problem statement.

VIII. Conclusion

The introductory programming (IP) course is considered as one of the difficult courses for novice programmers. The teaching materials used in the IP course focuses more on programming knowledge (syntax and semantic) of the programming language. Moreover, it promotes surface learning, and programming shortcut. This study re-designed the IP teaching materials based on the four stages of the ADRI model. The ADRI editor was prepared to support the four stages of the ADRI approach in the IP course.

The IP teaching materials based on the traditional and ADRI approaches were compared against five categories such as *Teaching topics for lecture sessions, Practical topics for lab sessions, Types of programming examples and problems, Presentation style of example and problems,* and *Four problem solving steps in examples and problems*. The comparison shows that the ADRI approach promotes both the programming knowledge (syntax and semantics) and problem solving strategies compared to the traditional approach who emphasizes more on programming knowledge. The traditional approach promotes programming shortcut presentation style (problem statement → Codes) whereas the ADRI approach reduces it by following presentation style (Problem Statement → Solution Plans → Codes) which promotes deep learning of the programming domain.

References

[1] I.M. Sohail, and C.N. Jo, "Impact of a new teaching and learning approach in an introductory programming course, Journal of Educational Computing Research, SAGE, vol. 55, No. 6, pp. 789-819, 2017, DOI: 10.1177/0735633116685852 (2017)

[2] S. Iqbal, and O.K. Harsh, "A self review and external review model for teaching and assessing novice programmers", International Journal of Information and Education Technology, Vol. 3, No. 2, 2013, pp. 120-123

[3] C.C. Wu, J.M.C. Lin, and K.Y. Lin, "A content analysis of programming examples in high school computer textbooks in Taiwan", Journal of Computers in Mathematics and Science Teaching, Vol. 18, No. 3, 1999, pp. 225–244

[4] S. Mohorovicic, and V. Strcic, "An overview of computer programming teaching methods", Proceedings of Central European Conference on Information and Intelligent Systems, CECIIS, Croatia, 2011, pp. 47-52

[5] Hu, "A case study in teaching adult students computer programming", Proceedings of the NACCQ conference, NZ, 2003, pp. 287-290

[6] A. Gomes, and A. Mendes, "A teacher's view about introductory programming teaching and learning: difficulties, strategies and motivations", proceedings of IEEE Frontier in Education (FIE), Madrid, 22-25 Oct, 2014, 2014, pp. 1-8

[7] L.E. Winslow, "Programming pedagogy—a psychological overview", ACM SIGCSE Bulletin, Vol. 28, No. 3, 1996, pp.17-22.

[8] I. Sohail, M. Roy, C. Morshed and C.N. Jo, "Impact of Assurance of Learning in programming course for novices", Proceedings of the 2nd British Computer Society (BCS) International IT Conference, Abu Dhabi, UAE. 2014, DRO link: *http://dro.deakin.edu.au/view/DU:30073423*

[9] S. Iqbal, M. Chowdhury, and O.K. Harsh, "Mobile devices supported learning for novice programmers", Proceeding of the 2nd international conference on E-learning and E-technologies in education, IEEE, Poland, 2013, pp. 277-282

[10] F.J. Tapia-Moreno, H.A. Villa-Martinez, C.A. Lopez-Miranda "Elaboration of statistics learning objects for mobile devices", International Journal of Interactive Mobile Technologies, 2012, Vol. 6, No. 2.

[11] D. Mcphee, P. Thomas, M. Ware, and N. Thomas, "An investigation into the effectiveness of the use of m-learning in the teaching of multimedia to first year university students", International Journal of Emerging Technologies in Learning, 2006, Vol. 1, No. 1.

[12] E. Abdalla, S. Mohanaad, Y.A. Mohamad, I.M. Sohail, "Using Information Communication Technology as a Teaching tool in Sudanese Governmental Universities of Khartoum State", International Journal on Informatics Visualization, 2017, Vol. 1, No. 4, Pages. 150-156

[13] P. Gurunathan, M. John, I. Sohail, "Bringing Technology into the classroom: Receptivity and Satisfaction level of faculty members Using Smart board – An Experimental Study" In proceeding of ICAICT 2014, Elsevier, pp. 274-279.

[14] ACM-IEEE Joint Task Force on Computing Curricula 2013, Computer Science Curricula 2013, ACM Press and IEEE Computer Society press DOI: http://dx.doi.org/10.1145/2534860.

[15] J. Tavares, I. Brzezinski, I. Huet, A. Cabreal, and D. Neri, "Having coffee with professors and students to talk about higher education pedagogy and academic success", Proceedings of the 24th International HERDSA conference, 2001, Newcastle, Australia

[16] M. De Raadt, M. Toleman, and R. Watson, , Textbooks Under Inspection, University of Southern Queensland, Australia, 2005, retrieved May 2013, <http://eprints.usq.edu.au/167/1/TechReport_Draft_10.pdf>

[17] I.S. Malik,. "Enhancing practice and achievement in introductory programming using an ADRI editor", In Proceedings of the IEEE International Conference on Teaching, Assessment and Learning for Engineering, 2016, pp. 32-39, IEEE, Thailand, 7-9 December, DOI: 10.1109/TALE.2016.7851766

[18] I.S. Malik, and J. Coldwell-Neilson, "Comparison of traditional and ADRI based teaching approaches in an introductory programming course". Journal of Information Technology Education: Research, 16, pp. 267-283. 2017, http://www.informingscience.org/Publications/3793

[19] T. Jenkins, "Teaching Programming – A Journey from Teacher to Motivator", Proceedings of 2nd Annual LTSN-ICS Conference, London, 2002, pp.65-71.

[20] I.S. Malik. "Role of ADRI model in teaching and assessing novice programmers", PhD thesis, Deakin University, 2016, http://hdl.handle.net/10536/DRO/DU:30088862

[21] M. Webster, Overview of programming and problem solving, Merriam-Webster's Collegiate Dictionary, Tenth Edition, 1994, retrieved on July15, 2013: computerscience.jbpub.com/vbnet/pdfs/mcmillan01.pdf

[22] I.M. Sohail, and C.N. Jo, "A model for teaching an introductory programming course using ADRI", Education and Information Technologies, Springer, vol 22, issue 3, pp. 1089–1120, DOI: 10.1007/s10639-016-9474-0

[23] I.S. Malik, "Improvements in introductory programming course: action reseach insights and outcomes", Systemic Practice and Action Research, DOI: 10.1007/s11213-018-9446-y, 2018, Springer

Use of Cloud computing for content delivery via 'Digital Ecosystem'

Swapnil Morande

Majan University College, Muscat, Sultanate of Oman

Dr. Veena Tewari

Majan University College, Muscat, Sultanate of Oman

Abstract: 'Digital Ecosystem' is the interdependence of computer software, hardware and digital services that play a role to fulfil organization's objectives. It is a convergence of recent broadband adoption and technological advances that has pushed the internet, telecommunication technologies and media industry into a flux; creating a scope for greater market penetration. One of the biggest advantages of Digital Ecosystem include mapping of consumer interest to gain competitive advantage in the market.

As of today, digital firms have been designing state of the art hardware - some of them are working to make software while some are creating value based digital services. In fact, without Digital Ecosystem it is not possible to fulfil the promises to the consumer and be the leader in any industry. The research presents various tangible and intangible benefits received by the organization with the help of Digital Ecosystem. The proposed research will further discuss the components for enhanced user experience. It is to be noted that many organizations are yet to implement a respective Digital Ecosystem for business. Hence, researchers plan to provide an outlook on how technology works hand in hand with the products and services.

In the future, customers will align their decision making based on such 'Digital Ecosystems' they intend to be part of. The customers will also be influenced to buy the products/services based on the ecosystem that suits best for their needs and lifestyle. The variables involved in this research are complex as well as difficult to measure, hence the qualitative approach has been followed to integrate various elements to devise a business strategy.

This research is significant because 'Digital Ecosystem' can directly or indirectly influence profit margins of a business and it can further be integrated as part of business strategy that can push the products and services as well as the content based on their consumption pattern of a customer.

Keywords: *Digital Ecosystem, Content Management, User Experience, Cloud Computing, Service Ecosystem.*

1. Introduction

The metaphor of 'Digital Ecosystem' describes the use of systems and innovative technological practices to achieve customer satisfaction with greater social inclusion. The research comments on the value addition that an organization can have while utilizing such Digital Ecosystem. Proposed research also takes into consideration the current Cloud computing practices and evaluates a platform for enhanced productivity. The assumption behind Digital Ecosystem is, that it functions as a whole. Based on the design of the hardware, the function of the software interface and the services (in the form of content delivery), ecosystems can target a specific set of customers and can lock the customer down (customer loyalty) in the market. Another fact about Digital Ecosystem is that it creates dependencies across product/services and then services (i.e. Content). These dependencies can further lead to greater consumption of the content. This research provides a new outlook on how technological advancements – in the form of Cloud computing - which pushes the boundaries of customer experience to its limits.

2. The research question

'Digital Ecosystem' coupled with products and services is making content delivery a lot more appealing to customers.

Hence proposed research will work towards following-

- Understand the role of 'Digital Ecosystem' in an organization.
- Identify the extent of the influence of 'Digital Ecosystem' on an organization.
- Investigate business opportunities presented via 'Digital Ecosystem'.

3. Justification for the proposed research

Now a days it is not enough to be able to sell a product or services and expect customers to become loyal without providing value added services. This research will showcase how an organization can leverage different components of a Digital Ecosystem in a holistic manner. It will also demonstrate how content in the form of digital delivery can help companies achieve a market leadership.

Various attempts have been taken to explore and think about business models and ecosystem as separate entities. But there are very few studies which have focused on their combined exploration. By studying the Amazon's (B2B & B2C) business models and ecosystem analysis, we have tried to understand how various business models relate to quality parameters in today's service industry.

4. Research methodology

The objective of the research is to explore the extent of the influence of 'Digital Ecosystem' on the core content delivery by any service provider. Researchers have carried out a conceptual study of Digital Ecosystem. This helped in developing an initial understanding of the concepts and finally making relevant observations to present researcher's perspective. Researchers have studied the different applications and followed up with advancements in the area of E-Commerce (B2B and B2C models). They have further reviewed various publications since a decade to identify and develop the initial and contemporary variables of the area in focus.

The primary research method of this study was a qualitative which is one of the best approaches to increase understanding of complex social constructions, such as ecosystems (Huberman, 1994). The methodology has been adopted to justify and comment on the objectives. These justifications and explanations have been used to understand the role of Digital Ecosystem in any organization and further to quantify the tangible and non-tangible benefits of a Digital Ecosystem.

5. Literature review

Digital technology, helps service providers, organizations and content publishers, achieve an unsurpassed level of distribution reliability. In fact, technology – such as 'Cloud computing' - further lowers the cost of delivering rich content even for additional users. It increases the audiences' base with minimal network bandwidth and with server optimization. It further makes it possible to have the flat-cost model of traditional broadcasting with the reliability and scalability as well as access to the Internet. Overall, it offers a cost effective way to deliver rich content in a service ecosystem.

In a service ecosystem the customer experience gets enriched by providing standard solution that integrates with cross platform solutions for the delivery of rich content. (High-Tech Writers, 2001) With the technological advancement, companies have developed innovative practices by recognizing that software needs to be coupled with hardware platforms. This gives rise to the concept of 'Digital Ecosystem' that represents connected digital devices and their environment functioning as a whole. (Bajarin, 2011) It can be mutually exclusive of service ecosystem offered via Cloud based services.

The Digital Ecosystem can combine websites, smartphones and desktop application with various online services where businesses can reflect their operational strategy. (McCormack, 2011). As per Forbes (2015) as technology continues to progress exponentially - Mobile, Cloud based technologies are driving a new generation of digital businesses. That said organizations are stretching their boundaries by identifying various practices within e-commerce to develop 'Digital Ecosystems' from the ground up. According to Briscoe and Marinos (2009), 'Digital Ecosystems' are distributed adaptive open sociotechnical systems, that are inspired by natural ecosystems and appears to be a novel approach for enterprises. The research states that 'Digital Ecosystems' can further aim to help small business entities become active players in globalisation and create value networks. It can be inferred that this approach can be considered as an effective strategy of globalisation that preserves regional growth and identity and can be embraced within several e-commerce models.

Thor Olavsrud (2015) documents that organizations are working with manufacturers to make unified communication work across devices using Internet of Things. Some organizations are working to create a platform through which an ecosystem of applications would help enable collaboration and workflow between stakeholders (Product Manufacturer, Service provider and Consumers). In his article, he further notes – ''As the digital transformation continues, today's organizations are building Digital Ecosystems that tap an array of other digital businesses, digital customers and digital devices at the edge of their networks.'' This research will further comment on the active and functioning Digital Ecosystem in an enterprise and it will further evaluate that ecosystem in terms of its processes and also in a holistic manner. This research will also elaborate how some unique ecosystems can help organizations keep the customers more loyal by offering an enhanced user experience.

As per the article published by McKinsey (2017), such re-engineering of processes by adopting an ecosystem view can help capture external data and provide analytics based services to attract more customers. A technology with the help of applications and backward-compatibility of evolving product lines can provide additional capabilities to provide an enhanced experience to create 'Network effects' that leverages the control of core technologies. (Farrell and Klemperer, 2006). 'Network effects' is nothing but the increase in the overall value of the product or a service when more people start using it. According to Reddy (2018) Network effects help scale a business by increasing customer base, market share and the overall value proposition of the product/service or content. Such Digital Ecosystem can integrate multiple services to holistically address customers' needs, making product and services further attractive. It can also help deliver the products and services as a bundle. By using Digital Ecosystems, organizations can monitor entire networks leverage this access to get new customers, tap into new sources of data, and improve existing business processes. (Desmet, Maerkedahl and Shi, 2017)

The principle of Digital Ecosystem is capable of transforming other sectors such as Oil and Gas industry. (Santamarta, Singh and Forbes, 2017.) This digitalisation brings in a different perspective towards value chain and redefines roles as per new business models to be able to achieve competitive advantage. According to the Accenture Technology Vision report (2015), modern organization will no longer act as isolated entities, but component pieces of new Digital Ecosystems in order to tap a broad range of businesses, customers using networked devices and technological advancement.

According to HÄIKIÖ and KOIVUMÄKI (2016) there are multiple factors originating from information technology, process and business related which may affect value creation during the digital service innovation process. The most important role is of information technology, which is multifaceted, providing both new opportunities and challenges in any service ecosystem. Present research is looking forward to explore the same.

6. Theoretical framework

The research includes the study carried on Amazon Inc. where researchers have observed and evaluated its E-commerce driven framework. Amazon has been considered for this research due to its capabilities in cloud computing, competencies in content management and strategic fulfilment achieved via Digital Ecosystem. (deAgonia, 2013.) Amazon has been strong in enabling its e-commerce platform and working to strengthen its strategy as an organization. The company has been investing in hardware and software component of ecosystems in order to differentiate and gain competitive advantage.

In this new era of truly personal computing, being led by devices like smartphones, and now tablet PCs, Amazon has introduced its Fire Series products. It is to be noted that amazon made an attempt to bring in another hardware Amazon fire Phone which resulted in a failure, but ended up giving insights to the company regarding hardware development. (Mangalindan, 2014.) Those insights reflect in 'Amazon Fire Television' where it works with several manufactures in order to create an open ecosystem and help the company make more money via content delivery.
When it comes to customized software, although Amazon has modified the existing Operating System (OS), a certain number of core applications are expected to get launched utilizing cloud based services to add more value to their ecosystem.

The mechanics of Amazon's Ecosystem, revolves around Kindle devices. It is a hardware that comes in various versions and bundled with various digital services. Amazon works around overall shopping experience and personalized content delivery leveraging its content based platform.

The E-commerce activities that run through its portal display a stronghold of Amazon as shown in the given figure 1.1.

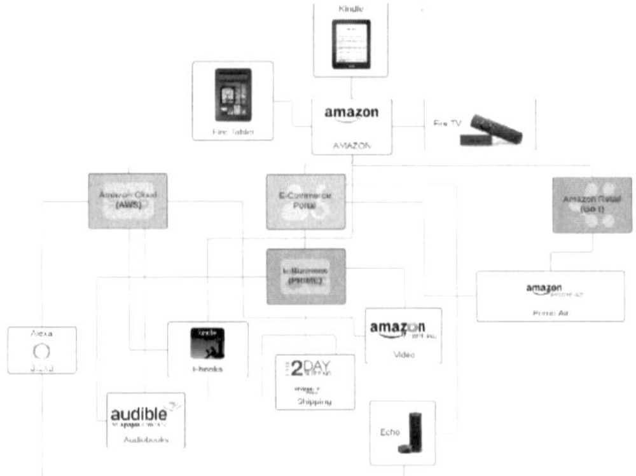

Figure 1.1 Digital Ecosystem - Amazon Inc.

From the above block diagram (Figure 1.1), one can identify the most important elements such as Kindle, Fire TV etc., coupled with e-commerce services.

The Cloud business of Amazon includes Amazon Web services (AWS) as well as a thriving virtual assistant called 'Alexa'. The AWS further helps Amazon facilitate online videos as well as shipping services along with audio books and e-books as well. The cloud-based service - Alexa- houses three products named Amazon echo, Amazon plus, Amazon echo plus. These three products (with few more being introduced in the near future) interface with Amazon Alexa to fulfil various operations that involve getting groceries, checking weather, booking flights, calling radio cabs and getting a burger.

In the near future, that Amazon will monetize the content available on Alexa. There's a possibility that Amazon might sell advertisements on Alexa 's hardware devices as well.
It can also possible that amazon's business model will make money by providing access to third party services leveraging the existing Cloud platform and Digital Ecosystem.

Further, it can be noted that the content delivery is greatly enhanced for activities such as reading a book. This is where Amazon has the biggest store of e-books and paperback editions. With its Cloud based synchronization service (also known as 'Whisper sync') it lets reader sync progress as a bookmark across devices and carry out reading on the move. Amazon can also play the media (Music and Radio streams) – via 'Alexa' - where the store makes use of cloud player to fulfil the content consumption and deliver music to its own devices.

As a matter fact, Amazon is the undisputed king of online shopping, where it sells selling almost everything. With its development of hardware and software Amazon has certainly made its services platform independent, at the same time ecosystem pushing customers across various amazon based services. One of the very know example of this is Amazon Prime where its two-day shipping service provides value to all the Kindle device owners by delivery free eBooks as well as Content in the form of online Music and Movies. (Broida, 2017) So far the company has managed to maintain better integration across all its components as it can be seen in the Digital Ecosystem. To be able to push the available content. In the present age of consumerism; Amazon has developed itself as one of the most proactive service providers. Coupled with the importance of bringing tangibility in any service ecosystem the importance of Cloud computing and Internet of Things (IoT.) It does not stop there, as of 2017 the nascent retail business Amazon is making use of Artificial Intelligence in order to develop a retail experience with 'Just Walk Out' technology to fulfil the promise of the contactless shopping system via Amazon Go (Stark, 2017.)

The proposed model (Figure 1.2), is an attempt to integrate the benefits of using above mentioned technologies for content delivery in the service sector by any contemporary service provider. The model not only takes care of core service delivery, but also covers the importance of facilitation of payment options. The model can act as a driver to enhance the customer's overall joy of owning any service/product. This can be followed and practiced by any service provider to deliver never before customer experience by directly supporting service quality parameters as shown in the framework below.

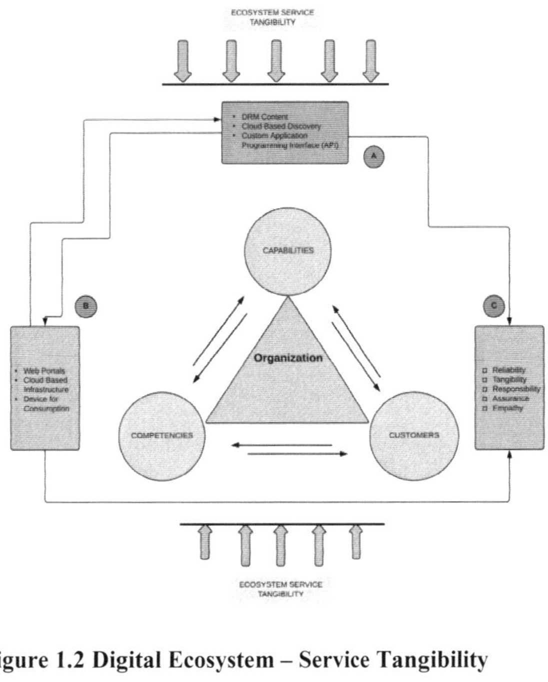

Figure 1.2 Digital Ecosystem – Service Tangibility

During this research, researchers have referred the Amazon's model, as a base to develop the proposed framework because it is one of the pioneer brands that plays to exploit the capabilities. These capabilities are used in terms of its an acquired resource - not only to refine the competencies of its human resources but also the process oriented competencies. This relates to the basic assumption of service marketing of having 3 extended P's i.e. People, Process and Physical evidence to manage any service/product. Today's service provider will be able to manage its vast portfolio of softwares, hardware and services by realigning its resources using the model shown above for any type of content delivery (Figure 1.2).

7. Contribution of the research

The contest between various organizations has evolved into locking the consumers down by leveraging online services along with hardware and softwares. The research contributes to the significance of alignment of these three components. This research helps organizations gain a bird's eye view on the holistic concept and essential components of Digital Ecosystem. Thus, this research explains a specific Digital Ecosystem, that uses using existing technologies with the objective of creating net value and building a competitive infrastructure.

8. Conclusion

Business strategies have come a long way where value added services were limited. Such a product/service paradigm has changed completely in modern times, and more efforts from the manufacturer or service provider are expected to retain a customer. The existing competition among companies is increasingly moving to the combination of hardware, software and services. This effectively gives rise to realm of Digital Ecosystem. Where an organization can make use of evolving technologies aligned with Digital Ecosystem to deliver betterment of consumer experience. In today's e-commerce driven environment where a huge rush of facilitating services to provide enhanced customer satisfaction, given framework can be exploited within any service ecosystem. On the basis of a review of literature, observation and analysis, technology can be used both as a mediator and final service product for contemporary service industry. If it is being utilized effectively and in an evolved way, can create competitive advantage by enhancing quality parameters in any e-commerce business model. This can be achieved by providing different value propositions to different stakeholder groups within the service ecosystem.

References

Anonymous 2012, Jan 30. *Mobile Cloud Computing Industry Outlook Report*: 2011-2016. *PR Newswire*.

Anonymous 2012, Oct 24. *Dell Extends ARM-based Server Ecosystem Enablement with Open Source Development for the Apache Community*. *Business Wire*.

Bajarin, B. (2011). Why It's All About the Digital Ecosystem. *[online] Tech.pinions. Available at: https://techpinions.com/why-its-all-about-the-ecosystem/4567 [Accessed 6 Jan. 2018].*

Briscoe, Gerard, and Alexandros Marinos. (2009). "*Digital Ecosystems in the clouds: towards community cloud computing*." Digital Ecosystems and Technologies, 2009. DEST'09. 3rd IEEE International Conference on. IEEE, 2009.

Broida, R. (2017). *Here's everything you get with Amazon Prime*. [online] CNET. Available at: https://www.cnet.com/how-to/amazon-prime-20-benefits-every-member-gets/ [Accessed 10 Jan. 2018].

Chang, Elizabeth & West, Martin. (2006). *Digital Ecosystems a Next Generation of the Collaborative Environment*. 3-24.

Dempsey, L. (2018). *Decentred network presence*. [online] Lorcan Dempsey's Weblog. Available at: http://orweblog.oclc.org/decentred-network-presence/ [Accessed 9 Jan. 2018].

Desmet, D., Maerkedahl, N. and Shi, P. (2017). Adopting an ecosystem view of business technology. [online] www.mckinsey.com/. Available at: https://www.mckinsey.com/business-functions/digital-mckinsey/our-insights/adopting-an-ecosystem-view-of-business-technology *[Accessed 13 Mar. 2018].*

Dwivedi, Y. and Mustafee, N. (2010). It's unwritten in the Cloud: the technology enablers for realising the promise of Cloud Computing | Journal of Enterprise Information Management | Vol 23, No 6. *[online] Doi.org. Available at: https://doi.org/10.1108/17410391011088583 [Accessed 6 Jan. 2018].*

EDITORS, B. (2001). *Digital Fountain Advances New Era of Rich Content Delivery through Partnerships with Industry Leader*s. Businesswire.com.

Eggleston, K. (2017). *Why the Digital Ecosystem is the Future of Business*. [online] ChaiOne. Available at: https://chaione.com/blog/digital-ecosystem-future-business/ [Accessed 13 Mar. 2018].

Farrell J. and Klemperer P. (2006), *Coordination and Lock-In: Competition with Switching Costs and Network Effects* . Available at SSRN: https://ssrn.com/abstract=917785 or http://dx.doi.org /10.2139/ssrn.917785

HÄIKIÖ, J. and KOIVUMÄKI, T., 2016. *Exploring Digital Service Innovation Process Through Value Creation.* Journal of Innovation Management, 4(2), pp. 96-124.

JANSSON, N., AHOKANGAS, P., IIVARI, M., PERÄLÄ-HEAPE, M. and SALO, S., 2014. *The Competitive Advantage of an Ecosystemic Business Model*: The Case of OuluHealth. Interdisciplinary Studies Journal, 3(4), pp. 282-295.

Mangalindan, J. (2014). *Why Amazon's Fire phone failed.* [online] Fortune. Available at: http://fortune.com/2014/09/29/why-amazons-fire-phone-failed/ [Accessed 10 Jan. 2018].

McCormack, R. (2011). *Digital Ecosystems: A framework for online business*. [online] http://bitstrategist.com. Available at: http://bitstrategist.com/2011/06/digital-ecosystems-a-framework-for-online-business/ [Accessed 13 Mar. 2018].

Michael deAgonia, J. (2013). *Battle of the media ecosystems: Amazon, Apple, Google and Microsoft*. [online] Computerworld. Available at: https://www.computerworld.com/article/2483616/personal-technology/battle-of-the-media-ecosystems-amazon-apple-google-and-microsoft.html [Accessed 8 Jan. 2018].

Olavsrud, T. (2015). How Digital Ecosystems are creating the 'we economy'. *[online] CIO. Available at: https://www.cio.com/article/2878419/innovation/how-digital-ecosystems-are-creating-the-we-economy.html [Accessed 6 Jan. 2018].*

Paul, D. (2015). [online] Forbes.com. Available at: http://The Rise of Digital Ecosystems in The "We Economy" [Accessed 6 Jan. 2018].

Reddy, N. (2018). *How to Harness the Power Of Network Effects*. [online] Forbes.com. Available at: https://www.forbes.com/sites/forbescoachescouncil/2018/01 /02/how-to-harness-the-power-of-network-effects/3/#3955061c1fad [Accessed 13 Mar. 2018].

Santamarta, S., Singh, R. and Forbes, P. (2017). *How Digital Will Transform the Upstream Oil Ecosystem*. [online] https://www.bcg.com. Available at: https://www.bcg.com/publications/2017/gas-technology-digital-transform-upstream-oil-ecosystem.aspx [Accessed 13 Mar. 2018].

Savov, V. (2012). *Digital Ecosystems: an in-depth comparison*. [online] The Verge. Available at: https://www.theverge.com/2012/4/18/2956951/ecosystem-comparison-amazon-apple-facebook-google-microsoft-sony [Accessed 9 Jan. 2018].

Stark, H. (2017). *Amazon Go, A Cashierless Convenience Store Now Available Near You.* [online] Forbes.com. Available at: https://www.forbes.com/sites/haroldstark/2017/11/20/introducing-amazon-go-a-cashierless-convenience-store-available-near-you/#788662e562f9 [Accessed 10 Jan. 2018].

Predictive modelling to illustrate factors influencing students at risk

Jacintha Menezes & Shajeea Khan

Majan College, Muscat, Oman

jacintha.menezes@majancollege.edu.om , **shajeea29@gmail.com**

Abstract

Students' academic performance and progression is the key performance indicator of any educational institution. Students' academic progression may vary due to various factors. Therefore identifying students at risk at an early stage could help to improve the academic progression of students though academic counselling and close monitoring. Predictive analytics can discover patterns in data that can lead to meaningful predictions for the unknown data. In the proposed research students already identified at risk are further categorised into low risk, medium risk and high risk. Student study level, Type of assessment, Type of module and Issues faced by the students are taken as the influencing factors. The research focused on finding the influencing factors with type of risk through predictive modelling. Research results show the issues faced by the student (Academic/Personal) as the main influencing factor with type of risk.

I Introduction

Data analytics has become the need of the day with data being stored digitally in vast amounts. Therefore it has become essential for organizations to use this data in order to get insight into the future. Since Educational institutions are also following the same trend of acquiring data, applying data analysis techniques will enable them to extract useful information [1]. Learning management systems have become part of majority of educational institutions which captures data useful to all the stakeholders.

The primary goal of any higher education is shape an individual students' personality by instilling in them the ethical values and state of the art skills. However students who enter higher education are of different capabilities in terms of motivation, interest and coping up with the subject arears. Therefore students' academic progression may vary and lead to dropout of non-progressing students. Identifying students falling in risk zone at an early stage could help the academic institutions to take corrective measures. In the proposed research an attempt has been made to see the influencing factors with type of risk which can used by the lectures/academic advisers to take appropriate actions to guide students at risk .

The remainder of this paper is organized as follows. Section II gives an insight on the related work already done in this field. Section III provides a brief introduction to predictive analytics and a few of the algorithms used for creating predictive models and Section IV presents the experimental results and discussion. Finally, Section V provides some concluding remarks.

II Related work

The field of education is one of the pioneer field to focus on for data mining procedures and knowledge discovery. M. Ghadavi and C. Patel [2] used predictive modelling and created a Linear

Regression based model to help predict the students' final exam marks based on their internal tests. González-Nucamendi, A. et al. [3] suggested the use of predictive analysis to find out the correlation between the students' final grades with their social and cognitive behaviour. A sample of 96 engineering students was taken and three different algorithm models were created to get the results using clustering, classification through decision trees and linear regression. The variables taken into account to predict the final results were Multiple Intelligence (MI), Self-Regulation (SR) skills and Learning Styles (LS). The results show that the knowledge of logical mathematics has a powerful effect on these students' final grades in a positive way while anxiety causes a negative impact on the grades. Moreover students with higher motivation levels were found to have gained better grades in the final assessment.

N. Kondo et al [4] used Learning Management System Log files to predict students who will be at risk at an early stage in the semester. The Log files taken over a period of five months contains the information on the frequency of the usage of the LMS by a particular student. This usage may be just log-in and log-off, to submit an assignment, to start e-learning etc. Logistic Regression was used for the predictive modelling with Support Vector Machine (SVM), Logistic Regression and Random Forest. The attendance rate of the students was found to be most important factor in prediction of the results and Random Forest algorithm gave the best prediction. Kaur, P. et al. used the classification algorithms to predict the students who are slow learners [5]. Different variables are used for this study which include the medium of instruction, the board, the gender of the student, students having digital devices etc. WEKA was used as an open source data mining tool and five classification algorithms J48, SMO, Multilayer Perception, Naïve Bayes and REPTree were applied. The comparison of the results obtained from all these classifiers shows that Multilayer Perception performs the best with 75% accuracy rate.

Research is done to assess the learning effectiveness of the students taking online courses [6]. The data taken was of 20 students and their online log files as well as their test data. WEKA software was used to perform C4.5 algorithm to produce a decision tree. The resultant decision tree showed that the important factors for students' achievements in an online learning environment was viewing the discussion boards, learning resources and logging in to the learning sessions. Nurbiha A. Shukor et al. concluded on the basis of this study that even though a student could be a silent learner in an online learning environment, however he/she can still achieve better results as long as they remain active in using online resources. Zhou, Q. et al. presented a research of students' performance of a course that was taken offline predicted based on their online behaviours [7]. The data taken for the experiments was the students' access records of the websites that they use. 200 thousand records from online websites and the student's grade scores from the course 'Data Structures' were taken as a test data set and Naïve Bayes Algorithm was performed to obtain the prediction results. The results obtained predicted more than 2/3rd of the data correctly.

In [8], Kumar, S. et al. used a decision tree to calculate student performance. The variables which are taken as the criteria to predict the performance were the students' skills which may be time management, problem-solving or online learning, motivation, self-learning etc. The predictive model built based on Multivariate Regression (M5P) which is a binary regression tree. The data set consisted of 307 students' from the Computer Science and Engineering department. Experiments performed included performance comparison between the M5P model and other classification algorithms J48, Naïve Bayes, AdaBoostM1 and RBF. The comparison showed that M5P performed better than other algorithms. Bharadwaj, B. and Pal, S. [9] used classification to predict performance improvement of students. Student related data with various variables was collected for 300 students and Naïve Bayes classification algorithm was applied. The results showed that the living area of a student has a direct impact on his/her performance. Moreover, the students' medium of study and their mother's qualification also have a high potential on their performance but it comes after the living location probability. This way the students that require attention on an urgent basis could be easily identified.

Data mining is also used to predict the performance of the instructor rather than the students which is also helpful for the educational institution. Four classification algorithms J48 Decision Tree, Multilayer Perception, Naïve Bayes and Sequential Minimal Optimization (SMO) [10] were used for the purpose of this study. The results show that the instructor's performance is affected based on the number of courses given to an instructor. Furthermore, it is observed that while J48 decision tree performs the best when the complete dataset is used among all four algorithms, SMO and Multilayer Perception performed the best when unneeded attributes were removed.

In the proposed research students already identified at risk are further categorised into low risk, medium risk and high risk. Student Level of study, Type of assessment, Type of module and Issues faced by the students are taken as the influencing factors. The research focuses on building a predictive model based on the influencing factor with type of risk.

III Predictive analytics

Data Analytics is the science of performing calculation on raw data for the purpose of receiving valuable information. The results obtained help businesses to acquire unforeseen knowledge helping them to make better decisions for the future. Predictive analysis in the field of data analytics focuses on providing future predictions based on the historical data that is already available. Predictive analysis develops a statistical model which helps in predicting future trends which in turn helps the organizations to be better prepared for it [1]. Data set that is already available is taken and different predictive analytics algorithms are applied the data set is trained after which the best model is selected to use on test datasets for predictions. Many different algorithms are available that can be used to create a predictive model. Classification and Regression algorithms are the two of the top-most techniques that are employed for this purpose.

Classification technique used in predictive analysis builds a model based on a training set. Test data fed to the model provides prediction based on the training data pattern [11]. Classification works by first determining the relationships between the variables leading to prediction of the output [12]. J48 algorithm is an open source implementation of the decision tree ID3 algorithm which results a binary tree. It splits the data into a range of values based on the values of the attribute provided at training phase. Random Tree algorithm can work for both classification and regression models. The splitting of the node is done based on a random selection of the data predicators [11].Reduced Error Pruning Tree (REPTree) is a fast learning decision tree algorithm. The logic used here is that of regression trees where multiple trees are generated. It generates decision tree based on information gain and minimizes the variance error [13]. MultiLayer Perceptron (MLP) is type of has multiple layers where hidden layers performs intermediate computation.

IV Experimental Results and Discussion

The records of student at different levels of risk was obtained from Registry department of Majan College. This dataset collected of over a period of six months and was prepared by selecting required input attributes. The selected attributes include level of study, Assessment and Module where the student had got a fail and issues faced by the student at risk. The following table gives the details of input and output variables and their possible values:

Variable Type	Variable	Possible values
Input	Level	{1, 2, 3}
Input	Type of Assessment	{Exam, Assignment }
Input	Type of Module	{ Theory, Practical }
Input	Issues	{Personal, Academic}
Output	Outcome	{Low Risk, Medium Risk, High Risk}

Table 1 Variable List

The prepared dataset was fed to Weka. The Weka is a freely available software which contains algorithms for data analysis and predictive modelling. In this research study, four different decision tree algorithms J48, Random Tree, MultiLayer Perception (MLP) and REPTree were used. All four of these algorithms generate a decision tree of the data provided by the user based on the different criteria that these algorithms work on. The comparative results generated by these algorithms provided in table 2,3 and 4.

Name of Classification Algorithm	TP Rate	FP Rate	Precision	Recall	F-Measure	MCC	ROC Area	PRC Area	Class
J48	0.714 0.687 0.902	0.070 0.054 0.216	0.606 0.891 0.793	0.714 0.687 0.902	0.656 0.776 0.844	0.601 0.673 0.688	0.879 0.829 0.867	0.515 0.747 0.781	Low Risk Medium Risk High Risk
Random Tree	0.643 0.795 0.892	0.032 0.092 0.180	0.750 0.846 0.820	0.643 0.795 0.892	0.692 0.820 0.854	0.652 0.711 0.712	0.948 0.932 0.945	0.746 0.889 0.927	Low Risk Medium Risk High Risk
REPTree	0.607 0.687 0.902	0.065 0.085 0.216	0.586 0.838 0.793	0.607 0.687 0.902	0.596 0.755 0.844	0.534 0.630 0.688	0.859 0.809 0.864	0.451 0.708 0.779	Low Risk Medium Risk High Risk
MultiLayer Perceptron (MLP)	0.714 0.807 0.824	0.070 0.115 0.126	0.606 0.817 0.857	0.714 0.807 0.824	0.656 0.812 0.840	0.601 0.693 0.699	0.896 0.924 0.937	0.693 0.884 0.917	Low Risk Medium Risk High Risk

Table 2 Detailed Accuracy by Class

TP - true positives are the number of outcomes predicted positive that are actually positive; FP - false positives are number of outcomes predicted positive that are actually negative. Precision provides fraction of the predicted positive are actually positive and Recall also referred to as sensitivity provides the outputs those that are actually positive were predicted positive.

Precision (P) = TP / (TP + FP)
Recall (R) = TP / (TP + FN)
F-measure is the weighted harmonic mean of Precision and Recall

F = 2PR/(P+R)

According to table 2 Random tree gives good precision and recall over other algorithms applied on the same data set. However F-measure shows outperformance of Random tree as compared to others. Moreover Random tree gives higher classification rate (82%) as compared to other algorithms as given below in table 3. REPTree gives the lowest classification rate (77%)

Name of Classification Algorithm	**Correctly Classified instances (%)**
J48	79.34
Random Tree	82.15
REPTree	77.93
MultiLayer Perceptron (MLP)	80.28

Table 3 Accuracy comparison

Root mean square error (RMSE) gives the average magnitude of the error which is the most commonly used error metric. Kappa statistic gives the measure of agreement of the prediction with the true class. Kappa statistic a more robust measure than simple percent agreement calculation. As per table 4 RASE of Random tree is lowest (0.28) and that of REPTree is the highest (0.34). Moreover kappa coefficient of Random Tree is the highest (0.7) and REPTree gives lowest coefficient (0.6)

Name of Classification Algorithm	**Root Mean Squared Error**	**Kappa Statistic**
J48	0.3292	0.6567
Random Tree	0.2817	0.6991
REPTree	0.3405	0.6303
MultiLayer Perceptron (MLP)	0.2951	0.6764

Table 4 comparison based on RMSE and Time taken

Based on the aforesaid results Random Tree outperforms as compared to other tree for the given data set. The decision tree model produced by the Random Tree algorithm is given below.

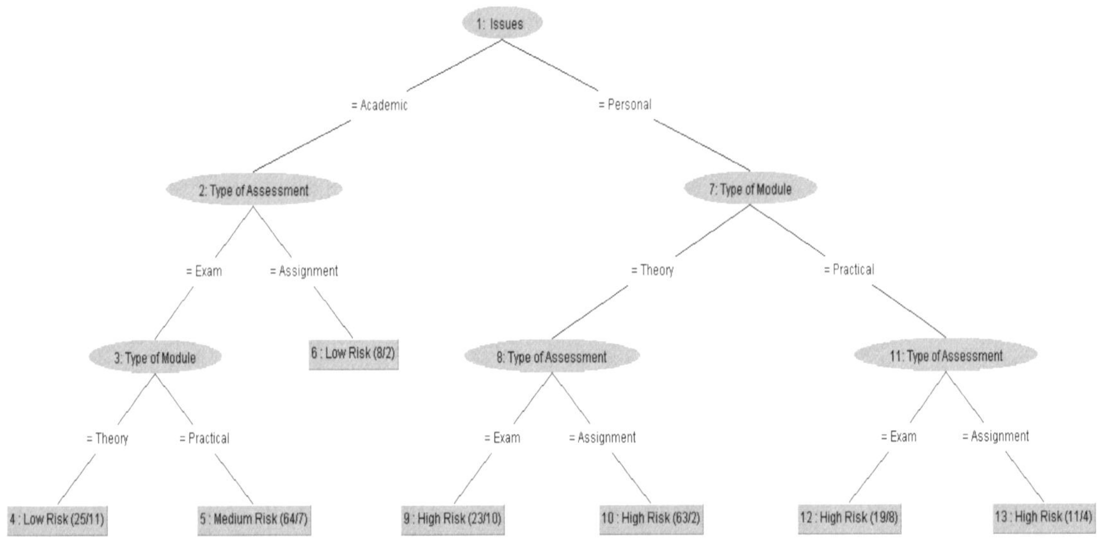

Fig 1: Decision tree model for the type of risks

Issues faced by a student considered to be major influencing factor with the type of risk. The issues faced a student are categorized as personal or academic. Personal issues covers health issues of self or family members, financial crisis faced by a student family, responsibility of student towards the family and family disputes. Student facing personal issues not necessarily be weak in studies rather they may fall into to the risk zone due to lack of continuous engagement with the module. Academic issues covers lack of specific skills required to complete a module. English language skills, research skills, problem solving skills, critical thinking skills, and subject specific skills are the common examples. Identification of the type of issue faced by a student at risk as a first step could help the academic advisor to guide appropriately.

According to the model shown (fig 1) academic issues are the main factor influencing student at low and medium risk which is linked to the type of assessment. The student with low risk could fail in an assessment due lack of module specific skills required. Since time module specific skills requirement is depicted in the module descriptor, necessary measures could be taken to fill the gap through workshops / in-class activities. In certain cases type of assessment is linked with type of module. If a module is of practical nature then good amount of lab work would be required to be put in by a student. Moreover practical modules build up the work progressively which requires student to be continually involved to be on the track. So there is a possibility of student falling into medium risk. Student with high risk facing personal problems could have issues with type of module which is linked to assessment due to non-engagement with the course of study. The Rule set generated by the tree can be described as follows:

IF Issues = Academic AND Type of Assessment = Exam AND Type of Module = Theory THEN Low Risk
IF Issues = Academic AND Type of Assessment = Exam AND Type of Module = Practical THEN Medium Risk
IF Issues = Academic AND Type of Assessment = Assignment THEN Low Risk
IF Issues = Personal AND Type of Module = Theory AND Type of Assessment = Exam THEN High

Risk
IF Issues = Personal AND Type of Module = Theory AND Type of Assessment = Assignment THEN High Risk
IF Issues = Personal AND Type of Module = Practical AND Type of Assessment = Exam THEN High Risk
IF Issues = Personal AND Type of Module = Practical AND Type of Assessment = Assignment THEN High Risk

V Conclusion.

The proposed reach made an attempt to find out the main influencing factor with types of risk. The dataset was subjected to different algorithms to build the decision model. Random Tree found to give comparatively better classification results. Based on the decision model Issues faced by a student at risk is considered as the major influencing factor. The study could help the academic institutions to identity and understand the students at risk at an early stage to take necessary preventive measures.

Bibliography

[1] B. G. Ephrem, Balasupramanian.N and H. Al-Shuaily, "Projection of Students' Exam Marks using Predictive Data Analytics," in *FREE AND OPEN SOURCE SOFTWARE CONFERENCE (FOSSC-2017)*, Muscat, 2017.

[2] M. Ghadavi and C. Patel, "STUDENT FINAL GRADE PREDICTION BASED ON LINEAR REGRESSION," *Indian Journal of Computer Science and Engineering (IJCSE),* vol. 8, no. 3, pp. 274-279, 2017.

[3] A. González-Nucamendi, J. Noguez, L. Neri and V. Robleda-Rella, "Predictive Models to Enhance Learning Based on Student Profiles Derived from Cognitive and Social Constructs," *IEEE,* pp. 5-12, 2015.

[4] N. Kondo , M. Okubo and T. Hatanaka, "Early Detection of At-Risk Students Using Machine Learning Based on LMS Log Data," *6th IIAI International Congress on Advanced Applied Informatics,* pp. 198-201, 2017.

[5] P. Kaur, M. Singh and G. S. Josan, "Classification and prediction based data mining algorithms to predict slow learners in education sector," in *3rd International Conference on Recent Trends in Computing 2015(ICRTC-2015)*, 2015.

[6] N. A. Shukor, Z. Tasir and H. V. d. Meijden, "An examination of online learning effectiveness using data mining," *Procedia - Social and Behavioral Sciences,* pp. 555-562, 2015.

[7] Q. Zhou, Y. Zheng and C. Mou, "Predicting Students' Performance of an Offline Course from Their Online Behaviors," *IEEE,* pp. 70-73, 2015.

[8] S. C. Kumar, E. D. Chowdary, V. S and K. V. K. Kishore , "M5P Model Tree in Predicting Student Performance : A Case Study," in *IEEE International Conference On Recent Trends In Electronics Information Communication Technology*, India, 2016.

[9] B. K. Bhardwaj and S. Pal, "Data Mining: A prediction for performance improvement using classification," *(IJCSIS) International Journal of Computer Science and Information Security,* vol. 9, no. 4, 2011.

[10] A. M. Ahmed, A. Rizaner and A. H. Ulusoy, "Using data mining to predict instructor performance," *Procedia Computer Science ,* p. 137 – 142 , 2016.

[11] S. Kalmegh, "Analysis of WEKA Data Mining Algorithm REPTree, Simple Cart and RandomTree for Classification of Indian News," *IJISET - International Journal of Innovative Science, Engineering & Technology,* vol. 2, no. 2, pp. 438-446, 2015.

[12] F. M. Ali, El-BahlulEmhemedFgee and Z. S. Zubi, "PREDICTING PERFORMANCE OF CLASSIFICATION ALGORITHMS," *International Journal of Computer Engineering and Technology (IJCET),* vol. 6, no. 2, pp. 19-28, 2015.

[13] L. Devasena C, "Comparative Analysis of Random Forest, REP Tree and J48 Classifiers for Credit Risk Prediction," *International Journal of Computer Applications,* pp. 30-36, 2014.

Int. J. of Technology Transfer and Commercialisation

MIC: Special Issue on, "Promoting Entrepreneurship and Technological Skills: National Needs, Global Trends"

Economic Benefits of Implementing Cloud in Higher Education Institutions, a reflection

Abstract: Most of the organizations have deployed communication and information systems that provide them seamless connectivity with partners and branches, and control on the resources, resulting in efficiency, easy management, time saving and enhanced productivity. As information and communication systems are becoming part and parcel of organizations without which modern enterprises cannot even think of existence, their reliability, security and cost effectiveness are some major concerns for enterprises to deal with. The adaption of cloud computing had been a revolutionary milestone in this regard. Higher education institutions have also benefitted from cloud implementations. The cloud implementations in higher education in Sultanate of Oman although being abysmally low however there is a gradual shift towards cloud hosting and implementation of clouds in higher education institutions. This research is a mere effort to highlight the economic benefits that higher education establishments can achieve by the implementation of cloud computing.

Keywords: Benefits of cloud, cloud computing, cloud advantages, cloud in HEI's, cloud and higher education, economic benefits.

(Anjum Zameer Bhat) '**Economic Benefits of implementing cloud in higher education institutions, a reflection**', *Int. J. of* Technology Transfer and Commercialisation

Biographical notes:

*This paper is a revised and expanded version of a paper entitled **[Economic Benefits of Implementation of Cloud in Higher Education Institutions in Sultanate of Oman]** presented at **[Majan International Conference, 20th March 2018]**.*

1 Introduction

Due to advancements in information and communications technology, we have witnessed a paradigm shift in how enterprises have evolved during past several years. Information and communications technology has emerged as an essential components of enterprises for success irrespective of the nature of work. As we observe more successful enterprises highly immersed in computers and information technology encouraging less penetrated enterprises to implement advanced means of information and communications technology. Due to various factors and changes in business world and impact of Information and communications technology, the essential aspects that can lead an organization to success have drastically changed in past few decades. We are living the age of communication, during past two decades we have witnessed the enormous development of communication and its impact on each and every sector. Organization of-late have realized that key to success in this competitive world is to immerse fully in to Information Technology and implement modern systems to assist their core business. Enterprises are continuously spending huge amount of money on information and communication systems to automate and facilitate promptness in their business process, increase customer satisfaction, ease of work, seamless operations, and control on business, improved communication, productivity and enhanced benefits. It is an evident fact that in past 2 decades organizations that have succeeded are those having improved and enhanced means of Information and communications technology implemented. Enterprise having enhanced systems are gaining cutting edge over their competitors and are able to perform better in all aspects compared to their counterparts having less technological aids available. We have seen boom of information systems of varying categories providing different features based on the business requirement of the organizations, implemented all over the world.

Higher education establishments also are not lagging behind as far as implementation of Information systems is concerned. We see educational establishments of every scale all over the world, implementing learning management systems, student information systems, content management and process management systems, virtual labs and virtual environments in addition to other communications technology to facilitate students and faculty members in teaching and learning. These modern aids of teaching and learning have significantly impacted the way of teaching and learning and has resulted in evolution of modern educational establishments. Although as complexity of information systems and learning management systems increases along with ever increasing demand for creation of new computer labs in higher educational institutions, an efficient, cost effective, viable, secure and feasible solution needs to be explored.

Cloud computing has brought revolution in the IT world by leveraging on-demand computing resources with 24x7 availability. Cloud computing provides computing as a utility alike any other utility like electricity, water, telephone and internet. Educational establishments all over the world have also utilized the cloud computing for hosting their IT resources and henceforth availing the features of scalability, availability, and pay-as-you-go and so on. Cloud computing offers variety of benefits in all its implementation forms i.e. public cloud, private cloud or a hybrid cloud to higher educational establishments. Although there are various reasons for educational establishments to restrain from implementation of cloud computing like security however various significant benefits like availability, reliability, scalability and economic benefits cannot be omitted as well. Cloud implementation of any form in higher educational institutions can yield enormous long term economic gains that is the need of the hour in this crunching economy. This research paper discusses the typical infrastructure of higher education establishments in Sultanate of Oman and economic gains of implementation of cloud computing in these institutions.

Typical infrastructure of higher education establishments in Sultanate of Oman consists of couple or more of information systems, learning management system, web portal, content management system, Human resource management system, communication systems apart from variety of different labs that are built to support various higher education courses. Most of the higher education institutions are equipped with general lab to cater the varying computing needs of the students. After doing a small analysis on the infrastructure of higher education institutes in Sultanate of Oman, some of the facts that came into picture are

i. Most of higher education institutions are having a private data center with an average of 20 servers, few have physical while some have virtual servers as well.
ii. IT team of at least 15 members on average.
iii. Software development team of at least 10 members who are responsible for implementation and development of various information systems and their management.
iv. High end hardware and software used to support various information systems including the learning management system.
v. Various types of labs established to support different students belonging to different courses.
vi. The hardware and software upgradation happens on an average within 4 years.
vii. Backing up of data related various systems is the responsibility of the core IT team or the information systems developer.
viii. High end hardware like System Area Network and Storage Area Network is used to take backups on various information systems.
ix. Restore point objective usually does not guarantee the complete data restore from the point failure.
x. Restore time objective is high in most of the information systems, implemented in house in higher education establishments.

Cloud computing would be a feasible and cost effective alternate to all the above problems that are faced by most of educational establishments in

Sultanate of Oman. There are enormous economic benefits for the implementation of cloud computing in higher education establishments that elaborated in the "Economic benefits" section.

Considering information retrieved from 7 higher education establishments viz-a-viz the infrastructure, all of these higher education institutions are having huge infrastructure for student labs. These lab are not only for computer science students however for other engineering branches as well, as a matter of fact as other engineering branches apart from various specializations in computer science, Information Technology and software engineering are also having some modules related to computer science and IT. Although there is a huge infrastructure that has been deployed for these labs however efficiency of these labs in terms of student flexibility of access is not so positive and satisfying. Students have lot of limitations in terms of accessing the resources of these labs and get a very limited time due to busy schedule of these labs (Bhat & Singh, 2017).

2 Literature Survey and Related Work

The Cloud computing is a developing innovative computing model for bringing computing facilities. This approach depend on numerous of prevailing technologies like, virtualization, Internet, Web services, and many others. The delivery of this facility in a pay-as-you-go manner largely through the Internet bounces this facilities a new uniqueness (Sultan, 2010). Yet for the educational institutions like the colleges and universities, the current inescapable growth, extraordinary digital speed of networks provides not merely access to additional well-organized computing rather a new proficient system and a prospect to reconsideration styles for providing IT related services. Globally various educational institutions are predominantly well situated to harvest the foreseen financial benefits of cloud computing prototypes, and such proficiencies are particularly longed-for in these enormously hard economic times (Waggener & Shelton, 2009).

While, it is fact that the cloud computing is here to stay, explicit roadmap for the expertise is expected to strengthen. At the same time the economic and business forces play a vital role to a rational inferences. The numerous computing applications by nature are general-purpose, hence offers remarkable economies of the scale provided the source can be unified. Though the academics and practitioners in the IT community are creating rapid advances in comprehending these assumptions in technical expressions, also it is believed that a corresponding argument needs to start from an economic perspective (Marston, et al., 2011). Also, there are studies whose results are promising and provide backing for the use of cloud solutions in the higher educational institutions (Mircea & Andreescu, 2011).

Similarly, the findings of a study revealed that the implementation of the cloud solution depends mainly on support of the top management, size of the educational institution, pressure from the competitors etc. (Low, et al., 2011). The organizations attain greater IT economies of scale with cloud solution model when they finance initially in managerial, relational, and technical capabilities (Garrison, et al., 2012). It is also observed by many that the cloud computing is growing rapidly with applications in education sector. At present, e-learning has become so powerful and popular trend. There are numerous educational institutions that can't have the funds for such investments and the cloud computing model is the unsurpassed solution, particularly in the universities where the usage of PCs are more rigorous (Bora & Ahmed, 2013). Nonetheless, the cloud computing is a significant model, with the prospect to ominously reduce the costs through optimization and augmented functioning and financial efficacies (Takabi, et al., 2010) & (Youseff, et al., 2008).

The Cloud computing can't be viewed as an independent phenomenon, rather it must be seen against the backdrop of financial and utilitarian realities facing the world. Also, there are many usages of cloud and numerous channels and mechanisms through which developing economies will benefit (Kshetri, 2010). Of course, there are a many factors that slow down the adoption of cloud computing, but there are more compelling reason and investigation suggests that decision takers should endorse as much as probable a rapid adoption of the cloud computing solution (Etro, 2011).

Additionally, a study on cloud computing benefits in education sector enlists them as providing scholastic resource storage and databases, e-mails, applications and tools for students and faculty. The main benefit is cost efficiency for the employment of the software and hardware. Also, this technology can improve excellence of present education system at an inexpensive cost (Ghazizadeh, 2012). The Knowledge Management is arriving an age where people are expected to make important influence to knowledge creation and management. The old Knowledge Management technologies proved to be expensive to implement and difficult to use. The new approach of cloud computing is anticipated to proclaimer a new era of knowledge at a much lesser cost (Sultan, 2013). Also, many analysis of results confirmed the straight result of cost savings on the comparative benefit of cloud computing as well as its secondary effect on cloud-computing adoption (Oliveira, et al., 2014). However, the adoption of cloud computing capacities various benefits to an organization, an effective adoption of the cloud solution in an organization, particularly in educational institutes necessitates an empathetic different dynamics and expertise in diverse domains (Masud, et al., 2012). The present economic conditions of Sultanate Oman may create a hurdle in the implementation of Cloud

Computing as the initial implementation cost is higher (Bhat, et al., 2017) however there are enormous benefits and higher education establishments need to think seriously for implementation of cloud.

3 Cloud implementation economic benefits

A typical higher education institution in Sultanate of Oman as indicated in the introduction section bears almost 4-6 information systems, learning management systems etc. In order to do management of these systems, the higher education establishment need to employee qualified engineers, developers and systems administrators. Apart from this, the cost that is incurred on software licenses and hardware upgrades is huge. The backup systems to save the precious data of the institution also are available at a very high cost. The administrative overhead and cost is also considerable when it comes to administration of these systems. Instead of spending huge of amount of money on infrastructure, manpower, hardware, software and so on still higher levels are reliability, fault tolerance, availability and manageability are not achieved. Moreover various higher education establishments are going for virtualization however this also incurs huge amount of expenditure as virtual environment vendor charges huge sums for the licenses and servicing.

One more area where higher education establishments are spending huge amount of money is the establishment of labs like software lab, hardware lab, network lab, project lab, database lab, multimedia lab and general lab. A typical higher education establishment has 15 to 20 established labs for variety of purposes in Sultanate of Oman. This incurs huge amount of expenditure not only on the maintenance of these labs however on the hardware and software. Moreover all these labs require upgradation of hardware and software after every three years which keeps the expenditure continuous for a higher education establishment. One more area of expenditure for HEI's is hardware and software for students.

The majority of the expenditures a higher education establishment bears for various IT resources can be reduced to a great extent by implementation of cloud computing. The information systems could be hosted on the cloud service thus eliminating the need for software and hardware upgradations for servers that support the respective information system, backup strategies and device, administrative overhead cost etc. As the service is hosted on the cloud, the availability is 24x7 with reliability and fault tolerance. The cloud implementation also provides dynamic scalability feature thus allowing higher educational establishments to choose appropriate services from the cloud eliminating the resource wastage.

The establishment of Virtual labs through cloud computing also yields enormous benefits for the educational establishments as there is no cost form establishing the lab. The establishment of virtual labs benefit the educational institution in multiple manners i.e. elimination of various labs those can be used for some other purpose, no licensing or hardware upgradation required, students the access the lab resources from literally anywhere. This is one more major area where cloud computing can provide huge economic benefits to higher education establishments. Moreover higher education institutions spend a huge amount annually for the maintenance of the physical labs, if these labs are eliminated, no expenditure will be done by the higher education institution.

One more important aspect where cloud services can be very beneficial for higher education establishment is the data storage for IOT implementations. The implementation of internet of things is inevitable in near future in higher education (Bhat & Ahmed, 2016) and cloud computing can hugely support the data availability, storage, access, analysis and visualization. The cloud computing cloud support the implementation of IOT in higher education in a great manner.

Cloud computing can also support the student centred teaching and learning and provide support e.g. for various types of student activities. Various applications and tools that are used for conducting the activities can be supported by cloud services hence eliminating any requirement for having these applications maintained in-house. As (Bhat, 2012) and (Bhat, 2013) suggest the use of various information technology tools for enhanced student centred learning. Facilitating game based and activity based learning can be to a large extent supported by the cloud implementations without much of the overhead for the educational establishments.

The expenditure done on the backup systems, data center and other costly infrastructure can also be eliminated by implementation of cloud computing. Moreover there are huge economic gains as far as software licenses are concerned. Most of the faculty members from various specializations and areas like civil engineering, mechanical engineer, computer science etc. are using very costly software applications. These applications are available on thousands of dollars and need to be renewed every 2 to 3 years. The implementation of cloud services would eliminated the need for such software licenses and can be accessed from cloud with pay-as-you-go. Many multimedia applications are also very costly and many higher education establishments are not able to afford such applications. Cloud implementation would facilitate the use of such applications on a very minimal expenditure.

A typical higher education institute in Sultanate of Oman can reduce its IT expenditure to 75% by implementation of cloud computing and hosting of various services on the cloud. These services include the hosting of information systems, learning management system, labs etc. This saved money can be utilized for some other developmental work.

4 Conclusion

Cloud computing is going to significantly influence the future IT strategy and infrastructure of organizations. Educational institutions especially higher education establishments are going to benefit from it hugely. As many of the services that are typically required in a higher education setup can easily be obtained through cloud with much more flexibility and features in addition to enormous economic benefits that are yielded by cloud implementations. Higher education establishments have also implemented cloud services however not to an extent as it should have been. Higher education institutions in Sultanate of Oman can hugely benefit in variety of aspects by implementing or adapting cloud computing. The economic benefits of cloud computing for higher education establishments are enormous apart from various other benefits cloud computing is providing. Cloud service adaption might be the future trend for higher education establishments especially in crunching economic conditions.

5 List of References

Bhat, A. Z., 2012. *BhatMeer Inspirational Model for Student-centered Teaching & Learning.* San Francisco, World Congress.

Bhat, A. Z., 2013. Inspiring Creative Minds. *Springer,* Volume 247, p. 137.

Bhat, A. Z. & Ahmed, I., 2016. *Big Data for institutional planning, decision support and academic excellence.* Muscat, IEEE.

Bhat, A. Z. & Singh, B., 2017. *Learning Resources as a Service(LraaS) for Higher Education Institutions in Sultanate of Oman.* Noida , s.n.

Bhat, A. Z., Singh, B. & Pandow, B., 2017. *Economic hurldes for the implementation of cloud in higher education establishments in Sultanate of Oman.* Dubai, s.n.

Bora, U. J. & Ahmed, M., 2013. E-Learning using Cloud Computing. *International Journal of Science and Modern Engineering,* 1(2).

Etro, F., 2011. *The economics of cloud computing.* Brussels, s.n.

Garrison, G., Kim, S. & Wakefield, R. L., 2012. Success factors for deploying cloud computing. *Communications of the ACM,* 55(9), pp. 62-68.

Ghazizadeh, A., 2012. *Cloud computing benefits and architecture in e-learning.* s.l., s.n.

Kshetri, N., 2010. *Cloud computing in developing economies.* s.l., s.n.

Low, C., Chen, Y. & Wu, M., 2011. Understanding the determinants of cloud computing adoption. *Industrial Management & Data Systems,* 111(7), pp. 1006-1023.

Marston, S. et al., 2011. Cloud computing — The business perspective. *Decision Support Systems,* 51(1), pp. 176-189.

Masud, M. A. H., Yong, J. & Huang, X., 2012. *Cloud computing for higher education: A roadmap.* s.l., s.n.

Mircea, M. & Andreescu, A. I., 2011. Using cloud computing in higher education: A strategy to improve agility in the current financial crisis. *Communications of the IBIMA.*

Oliveira, T., Thomas, M. & Espadanal, M., 2014. Assessing the determinants of cloud computing adoption: An analysis of the manufacturing and services sectors. *Information & Management,* 51(5), pp. 497-510.

Sultan, N., 2010. Cloud computing for education: A new dawn?. *International Journal of Information Management,* 30(2), pp. 109-116.

Sultan, N., 2013. Knowledge management in the age of cloud computing and Web 2.0: Experiencing the power of disruptive innovations. *International Journal of Information Management,* 33(1), p. 160–165.

Takabi, H., Joshi, J. B. & Ahn, G.-J., 2010. Security and privacy challenges in cloud computing environments. *IEEE Security & Privacy,* 8(6), pp. 24-31.

Waggener, B. W. & Shelton, 2009. Above-Campus Services: Shaping the Promise of Cloud Computing for Higher Education. *EDUCAUSE Review,* 44(6), pp. 1-6.

Youseff, L., Butrico, M. & Silva, D. D., 2008. *Toward a unified ontology of cloud computing.* s.l., s.n.

Smart Driving Trainer

Afrah Mohammed AL Hanzhali| .Dr.Reshmy Krishnan| Muscat College Student id:13043

Problem / Question

People in Oman facing problems in searching about driving training centres especially when they come from rural side of the country. They face difficulty to get good experience instructor for driving training. I try to solve this issue by making a mobile application for this so that any time, anywhere people can use it.

Hypothesis

- It saves time because instead of going long kilometers searching about driving training centers .
- It shows the location of the company through GPS.
- It shows experience instructors to the customers as per convenient time.
- Instructors are getting customers as per expected time and cost.

Project Overview

- This project is designed to serve people who want to get information about driving licence and availability of driving instructors in Muscat. Many people who come to work or study in Muscat from rural area of Oman feel difficulty in getting information about the driving learning details. This application can help such people since this contains information about instructors, location, convenient timing. Moreover instructors can register their timing and rate in the system so that customers can search for instructors in each location and fix the learning time.
- .

Variables / Research

- The idea of my project is to develop a mobile application to help people to find information about driving process. This project is done for Muscat Driving School (MDS) which is considered one of the big driving training centres in Muscat. This application focuses in Muscat Governorate because many jobs focus in Muscat. It includes information about instructor, price, time, date, contact number and other information.

Materials

Component	Minimum
Operating system	Windows 10
Technology Used	Android 3.0.1 ,php, NOX and MySQL
language used	java
Tools used	Android SDK
System type	64-bit Operating System
RAM	4.00 GB
Motherboard	Genuine Intel

Procedure

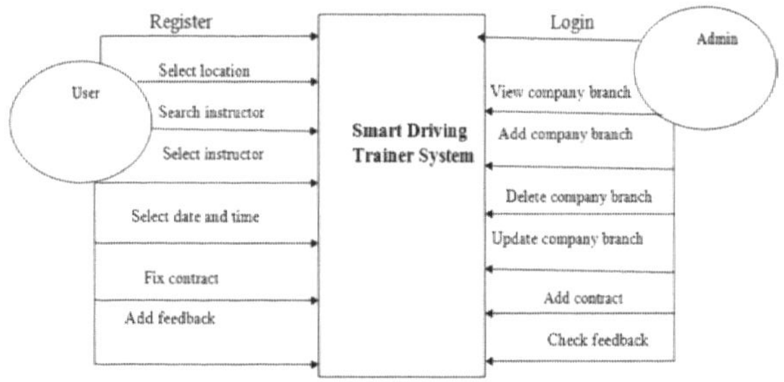

Data / Observations

- The user can register, login, select location, search instructor, select instructor, select date and time, search for terms and condition for price, rate company and fix appointment.
- The trainer can do register, login, search user and fix date and time. The trainer can login the system and search user to find address. Also, they can fix date and time with customer.
- The admin can do login, view company. Also, they can add new company. Next, they can delete company. Then, they can update company and check feedback.

Results

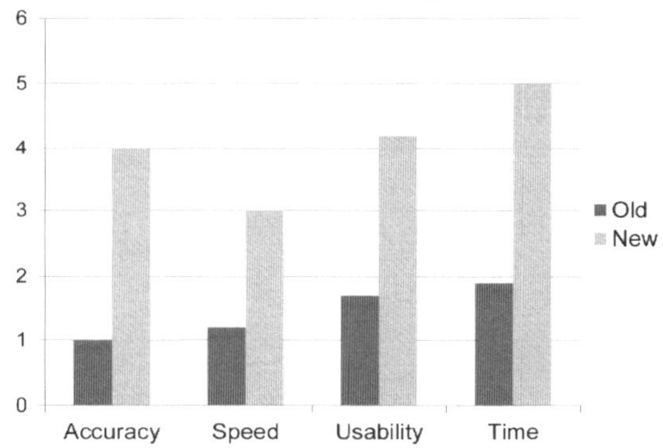

- This chart show comparison between past and today of Muscat driving school company for some point such as accuracy, speed, usability and time.

Conclusion

- To conclude, I have designed a mobile application for Muscat driving school to facilitate driving process. The application is called Smart Driving Trainer System. I have used android program to develop my application.
- .

Works Cited

- www.muscatdrivingschool.com/

Automated irrigation System

Maryam AL kaabi | Ali Alhumiri | Gutech

Abstract

The motivation for this project came from the fact that there are some countries with very little fresh water sources, and as we know, Oman is a water-scarce country, so it is necessary to develop such a smart system to help maintain fresh water sources without waste.

Introduction

Agriculture is one of the most important economic activities in the world because it provides food and goods for the world's population. The worldwide population is projected to reach 9 billion by the year 2050 (FAO, 2017).

Agricultural production should double to support the growing need for nourishment and bio-energy (FAO, 2017). Given the limited land, water, and labour, agricultural efficiency must increase by 25% to meet that target, all that while constraining the developing impact that farming has on the earth and its natural resources.

Project Overview

- Agriculture is one of the most important sectors of the country's economy, and it is our responsibility to take care of it. By combining technology with environments you may solve many issues. One of the big issues we face is that people do not tell their land that causes a lot of serious consequences. On the other hand, preserving plants may solve or reduce the consequences of many issues, such as global warming. The project has been driven by countries with little water. As we know that Oman is a rare water country, it is necessary to develop such a smart system to help conserve water without waste. In this project we have developed two different irrigation systems for trees in intelligent ways.

Objects

1. Increasing agricultural production using small quantities of water.
2. Minimize manual intervention in watering operations with increasing watering speed.
3. Preserving water from waste.
4. Reducing energy usage.
5. The ability to provide even distribution of water in the field.
6. Eradicate the human operating errors.
7. Reduce the number of workers in agriculture filed.

Materials

Materials (detailed list)	Quantity (be specific)
Temperature sensor	1
Moisture sensor	9
Rain sensor	1
Light sensor	1
Water level sensor	1
Valve	3
Pump	1

Pictures of the project

Data / Observations

- Observation 1
- Observation 2
- Observation 3

How the system works

This system is based on an intelligent system, which takes into account the amount of water used without waste. This process is done using different types of sensors. Furthermore, this system works in a complete way without requiring interference from humans. We are trying to make this project more intelligent by using other sensors connected to the system, for example, by stopping the irrigation process when there is sunlight and high temperature or when there is rain. Moreover, we have developed this project to be of high quality. For example, when there is a water leak, we can directly know this by using a flow-meter sensor, so that the system can be told to stop, to prevent wasted water. In the future, we will make this system work by using solar power.

Conclusion

- As mentioned, the field of agriculture is facing serious obligations in the future, as it should feed an exponentially increasing population, with limited amounts of natural resources. Automated irrigation and Robots are expected to play a major role in fulfilling the need of agriculture to do more with less, or in other words, improve its efficiency, by optimizing the use of natural resources like water and soil.

Reference

- A. Nickmanesh, M. R. (2011). Comparison between fixed and movable sprinkler irrigation systems. *21st International Congress on Irrigation and Drainage*, 349-361.
- Briscoe, J. (2009). Water, Agriculture, 1. and Development: The quality of advice? *Essays from the CSIS and SAIS Year of Water Conference, Washington*, 1-25.
- FAO. (2017). *The future of food and agriculture – Trends and challenges*. Rome.

Entrepreneurial competences in corporations - is there capacity for new ideas and opportunities?

Dushica Janevska[1], Gligor Stojkov[2], Radmil Polenakovik[3]

[1] FaSTO Ltd., Bul Ilinden 81/16, 1000 Skopje, Macedonia; [2] Deutsche Gesellschaft für Internationale Zusammenarbeit (GIZ) GmbH, Bonn, Germany, [3] Ss. Cyril and Methodius University, Faculty for mechanical engineering, Karposh II bb, 1000 Skopje, Macedonia

Abstract: (ABS) Contemporary corporations work in turbulent and dynamic business environment that requires high level of flexibility and capacity for recognition and utilisation of the new business opportunities. In order to reach this aim, the corporate managers should promote corporate entrepreneurship and acquire and cherish entrepreneurial competences related to development of new ideas and identification and deployment of new opportunities. Therefore, this paper presents a brief overview of the academic literature related to entrepreneurial competences, corporate entrepreneurship, the differences and similarities among the individual and corporate entrepreneurs as well as the barriers for introduction of corporate entrepreneurship. Additionally, this paper presents the findings of empirical research conducted among the managers in medium and large companies in the Republic of Macedonia. The aim of the research was to identify the level of presence of entrepreneurial competences related to development of new ideas and identification and utilisation of new opportunities on one side, and to explore the specifics among the managers belonging to different managerial levels on the other. Overall, high level of presence of the researched competences among managers was found, with larger number of managers showing strong presence of the entrepreneurial competences among middle and line managers compared to the top managers.

Keywords: (ABS) entrepreneurial competences, corporate entrepreneurship, managers

Reference to this paper should be made as follows: Janevska, D., Stojkov, G. and Polenakovik, R. (2018) 'Entrepreneurial competences in corporations - is there capacity for new ideas and opportunities?', *Int. J. Xxxxxx Xxxxxxx Xxxxxxx*, Vol. X, No. Y, pp.000–000. Harvard style

Biographical notes: (ABS)

1 Introduction

Contemporary companies are influenced by dynamic and continuously changing business environment. They are facing the challenges of the increasing competition, where everyone is trying to find a way to enlarge market share and improve the ability to generate income. In order to achieve these goals, companies need to increase their competitiveness and efficiency, to increase customer satisfaction, to improve service and product quality, to practice continuous improvement of employees, managers, services, products and processes, to act proactively in looking at latent customer needs and, last but not least, to be innovative and entrepreneurially oriented, while promoting new ideas and seizing new opportunities.

In this direction, the companies can undertake various activities. One of the key and most complex aspects to consider is to evaluate and continuously improve the competencies of the corporate managers in order to increase the capacity to adapt the company to the market demands.

2 Literature review

Entrepreneurial competences

Different authors accompany diverse competences with successful entrepreneurs, among them being: identification and utilisation of the new opportunities, creativity, innovation and risk taking. An entrepreneur is a person who creates a new company or investment (Gartner, 1988, Low and MacMillan, 1988) and is actively involved in the management of the company (Gartner et al., 1994). It is a person who recognizes an opportunity and creates a company for its implementation (Bygrave and Hofer, 1991). According to Baron (1998), the entrepreneur is capable of making changes, introducing a new way of doing things, exploring new opportunities and realizing ideas. For Ahmad and Hoffman (2007), the entrepreneur is a person seeking a way to generate value by creating or expanding economic activity through identifying and utilizing new products, processes, or markets. Among the authors who have identified the competence for creation, recognition and utilisation of the new opportunities as important trait of the successful entrepreneur are: Schumpeter (1934), Kirzner (1978), Drucker (1985), Buchanan and Vanberg (1991), Shane (2000).

The Joint research centre of the European commission developed the entrepreneurship competence framework – EntreComp as an effort of the European Union to develop common understanding of the entrepreneurship as a competence among its citizens (Bacigalupo et al., 2016). It aims to facilitate learning and exchanges between member states and to have a positive impact on the mobility, employment and active participation of citizens in society and the economy.

EntreComp defines entrepreneurship as universal competence, which applies to all areas of life: from personal development to active participation in society, competence required for entering the labour market or self-employment, but also for taking various actions in the sphere of culture, social life or business.

It is based on a broad definition of entrepreneurship as a capacity to act upon opportunities and ideas and transform them into cultural, social or economic value, thus covering various types of entrepreneurship, including corporate entrepreneurship. The EntreComp model consists of three areas of competences: "Ideas and Opportunities", "Resources" and "Into Action". Each area consists of five individual competences and all together are integral parts of entrepreneurship as overall competence. These are the entrepreneurial competences of the EntreComp model: Spotting opportunities, Creativity, Vision, Valuing ideas, Ethical & sustainable thinking, Self-awareness & self-efficacy, Motivation & perseverance, Mobilising resources, Financial & economic literacy, Mobilising others, Taking the initiative, Planning & management, Coping with ambiguity, uncertainty & risk, Working with others and Learning through experience (Bacigalupo et al., 2016).

Corporate entrepreneurship

The name corporate entrepreneurship is used to describe entrepreneurial behaviour in medium and large companies. It unites entrepreneurial efforts with company resources in order to create and realize innovative activities in the form of product, service, process or business model (Sathe, 2003).

The simplest forms of the concept of corporate entrepreneurship emerged in the 1970s as a result of certain changes in the market paradigm, the economy and the corporate culture. The roots of the corporate entrepreneurship can undoubtedly be found in entrepreneurship. The author who introduced the name corporate entrepreneurship (intrapreneurship) and corporate entrepreneur (intrapreneur) in 1978 is Gifford Pinchot. According to Pinchot, corporate entrepreneurship recognizes the following two aspects: (i) Corporate entrepreneurship is a set of good business practices that give full entitlement to persons in large companies that have entrepreneurial qualities to innovate quickly, not only for the sake of the company but also for the benefit of the customers, (ii) Corporate entrepreneurship encompasses individual and / or team activities with entrepreneurial behaviour, serving the interests of large companies and their supply chains, with or without official assistance (Pinchot, 2010).

Entrepreneurship is closely linked to individuals who, either in their own company or in the company they work for, use the opportunities independently of the resources under their direct control (Stevenson and Jarillo, 1990). Thus, a person can manifest entrepreneurial behaviour without having ownership or being shareholder of the company (Ahmad and Hoffman, 2007). Such behaviour is called corporate entrepreneurship or intrapreneurship.

Corporate entrepreneurship is defined as a new type of behaviour that includes intentions and activities other than the usual ways of doing business (Antoncic and Hisrich, 2001, 2003). This category plays an important role in the innovation and competitiveness of already existing and large companies (Pinchot, 1985, Zahra, 1991). For example, Pinchot emphasizes the need for generators of ideas and initiators in large companies and suggests that innovation does not occur without a dedicated and passionate small group of individuals, i.e. the corporate entrepreneurs (Pinchot, 1985).

In one of their publications on corporate entrepreneurship, Nielsen et al. (1985) suggest that corporate entrepreneurship is most appropriate for large corporations and is most applicable in a dynamic environment. Antoncic and Hisrich (2003) emphasize the following key elements of corporate entrepreneurship: new investments and businesses, product and service innovation, process innovation, risk-taking (potential losses, costs associated with new opportunities) and proactivity and consider them as tightly interconnected.

Ireland et al. (2009) emphasize the importance of the strategy for corporate entrepreneurship that the company should create. They developed a model of corporate entrepreneurship strategy, which consists of the following elements: (i) the foundation of a corporate entrepreneurship strategy (for example, individual knowledge of the conditions that support entrepreneurial activity); (ii) elements (for example, the entrepreneurial vision of the top management, the organizational structure that supports entrepreneurship), and (iii) organizational outcomes deriving from entrepreneurial activities. Thus, the strategy for corporate entrepreneurship can be defined as "the company's reliance on entrepreneurial behaviour guided by the company's vision, which deliberately and continuously leads to renewal of the company and forms the scope of its operation through recognition and utilization of entrepreneurial opportunities" (p. 21).

Similarities and differences between corporate and individual entrepreneurship and barriers for introduction

The definition of entrepreneurship as utilization of the opportunities independently of the available resources (Stevenson and Jarillo, 1990) applies both to corporate and individual entrepreneurship. Lukeš (2012) presents the similarities between corporate and individual entrepreneurship: both forms include opportunities recognition and the existence of a unique idea in the form of a new product, service, or process; understanding that the entrepreneur will encounter resistance and obstacles, which requires persistence to deal with them; the ability to formulate innovative solutions and development of creative strategies for optimal utilization of the resources. Finally, the two forms involve great uncertainty and require risk management strategies and an entrepreneur's ability to balance the vision with managerial skills, the passion for pragmatism and patience with pro-activity.

However, he further explains, there are differences between corporate and individual entrepreneurs. Corporate entrepreneurs carry smaller risks, and receive consequently smaller rewards as they are not owners of company's capital. As the large companies are not so susceptible to external influences as the smaller one, they provide more room for errors, thus protecting the individual. After all, large companies provide larger space for new ideas as well as better access to a variety of resources that allows fast business growth.

Certainly, entrepreneurial behaviour in large companies is followed by obstacles. For example, there is a great interdependence of corporate entrepreneurs with many other factors, such as rules, procedures and bureaucracy that make the decision-making process rather slow (Morris and Kuratko, 2002). In large corporations there are standardized processes and procedures that facilitate the efficient operation. Consequently, they focus on long-term planning, efficient resource utilization and development of experience based future steps. However, these conditions create comfort working environment that sometimes can decrease the motivation for making an attempt to introduce something new. This situation results in the creation of strategic, systemic, behavioural, as well as political barriers for introduction of corporate entrepreneurship (Lumpkin, 2007, Morris and Kuratko, 2002).

Managerial support of the corporate entrepreneurship

Managerial support refers to the managers' willingness to facilitate and promote entrepreneurial activity in the corporation (Kuratko et al., 2005). This support can take various forms, including support of the innovative ideas, provisioning of the necessary resources or expertise. Managers directly control and evaluate the work of their subordinates, allowing them certain level of access to the resources and information, clarifying the goals and objectives, giving an opinion on their ideas and influencing their work in many other ways. Managers play three important roles in supporting new ideas. They make an environment that supports the creation and communication of ideas. Secondly, they help employees to develop their knowledge and problem solving skills in order to increase the quality and the effect of the emerged ideas. And finally, managers "sell" the good ideas to the higher level managers and explore the possibilities for wider use of certain ideas throughout the whole corporation (Robinson and Schroeder, 2006).

Literature recognizes the support of managers as a factor that affects the creativity of employees (Amabile et al., 2004). Scott and Bruce (1994) confirm the positive role of the link between the leader and the team member in support of innovative behaviour. Kuratko et al. (2005) find that by increasing the entrepreneurial behaviour of managers, employees manifest greater satisfaction with their superiors, promoting them as supporters of entrepreneurial behaviour in corporations as well. Finally, Lukeš et al. (2009) show the importance of perceived managerial support as a factor of promoting innovation in corporations. In other words, even if the organization manifests positive attitude towards innovation, it cannot be implemented if the support from managers is missing. Therefore, support of the new ideas and innovation should receive due attention from top management, if the development of the corporation in the direction of entrepreneurial behaviour is desired.

3 Research aim

Having in mind the importance of presence of entrepreneurial competences among the managers in large corporations as a supporter and facilitator of the entrepreneurial corporate culture, which on the other hand is powerful toll for successful leadership of the contemporary companies that operates in dynamic environment, the research presented in this paper has two main objectives:

1. To identify the level of presence of entrepreneurial competences related to development of new ideas and identification and utilization of new opportunities among managers in medium and large companies in the Republic of Macedonia,

2. To identify whether among certain managerial levels there are differences in the presence of certain competences, aiming to understand which of the managerial levels can be the greatest supporter of the introduction and development of corporate entrepreneurship.

4 Methodology

For the purposes of this research, a quantitative approach was used, in accordance with the recommendations of scientific literature (Saunders et al., 2003). To this end, a self-administered questionnaire was developed, suitable for collection of the necessary data for answering the research questions. Beside the demographic data for the participants (age, gender, experience, education, managerial level, industry) the participants were asked to assess their attitudes towards 14 statements, aiming to measure the presence of the following entrepreneurial competences: *spotting opportunities, creativity, vision, valuing ideas and ethical and sustainable thinking.* The instrument was developed based on EntreComp model (Bacigalupo et al., 2016), particularly its first area "Ideas and opportunities", consisted of the above presented competences. Each statement was measured on a five level Likert scale (strongly disagree, disagree, neutral, agree and strongly agree), that in the data processing was coded as one to five integer values. The scales for measuring each of the targeted entrepreneurial competences proved to be reliable, resulting in Cronbach's Alpha values between 0,713 and 0,783.

Aiming to achieve high response rate, the invitation for participation in the research was sent to the managers to whom the researchers have direct access, in fact convenient sampling was applied. The target group was managers employed on managerial position in medium and large companies in the Republic of Macedonia who are not owners of the companies' capital. One hundred fifty seven corporate managers from different industries (information and communication technology, finance and insurance, manufacturing, transportation, different services etc.) were invited to participate in the research, out of which 112 have provided full and valid answers (Figure 1).

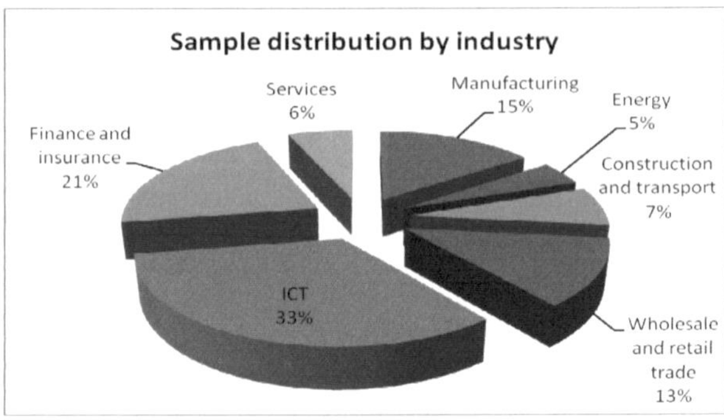

Figure 1. Sample distribution by industry

Thirty nine point three percent of the companies had between 50 and 250 and the rest 60,7%, have more than 250 employees. Sixty four point three percent were male participants while the rest thirty five point seven percent were female.

Thirty two percent were top managers, 33% middle and 35% were line managers (Figure 2).

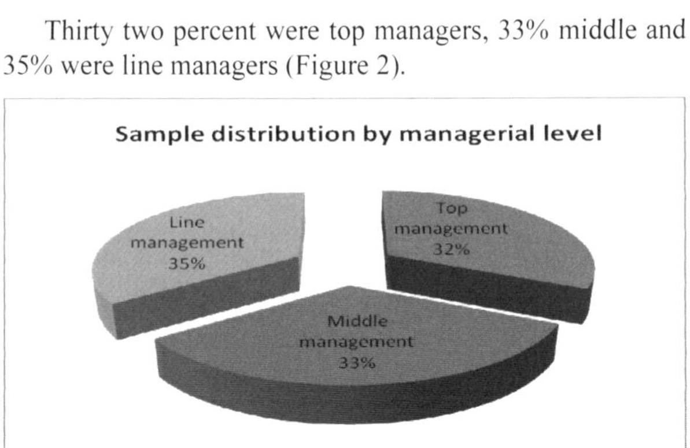

Figure 2. Sample distribution by managerial level

5 Findings and discussion

The main findings in terms of mean and standard deviation are presented in the Table 1, which illustrates the level of presence of each of the analyzed entrepreneurial competences among managers on the scale of one to five.

Table 1. Presence of entrepreneurial competences in area "Ideas and opportunities" among corporate managers

Ideas and opportunities	N	Mean	Std. Deviation
Spotting opportunities	112	3,99	0,4950
Creativity	112	3,96	0,4917
Vision	112	3,96	0,4785
Valuing ideas	112	4,04	0,5434
Ethical and sustainable thinking	112	4,29	0,5367

As it can be seen from the Table 1, relatively high presence of all competences is evidenced among the participants. The highest level of presence among the corporate managers, participants in the research, with mean value 4,29, has the competence *ethical and sustainable thinking.* It represents high level of ethical behaviour, sustainable thinking and accountable and responsible attitude among managers. They have capacity to assess consequences of the ideas that bring value and the effect of the entrepreneurial action on market, society and the environment. Additionally, the presence of this competence understands capacity to assess the sustainability of the long term goals and the course of the chosen action.

The other four competences with mean values around four, show that in the companies represented by the managers that participated in the research, there is a solid base for development of the entrepreneurial corporate culture that praise the new ideas and identification of the new opportunities, aiming to create new values.

In order to identify whether there are differences in the presence of certain competences among the managers belonging to different hierarchical levels, further data analysis were performed. Namely, the criteria "strongly present competence" was introduced, meaning that all participants that have level of presence of certain competence four or above satisfy the entrance criteria. Afterwards, the frequencies and percentual values for the managers with "strongly present competence" were calculated for each of the three hierarchical managerial levels: top, middle and line managers and each of the analysed entrepreneurial competences. The findings are presented on the Figure 3.

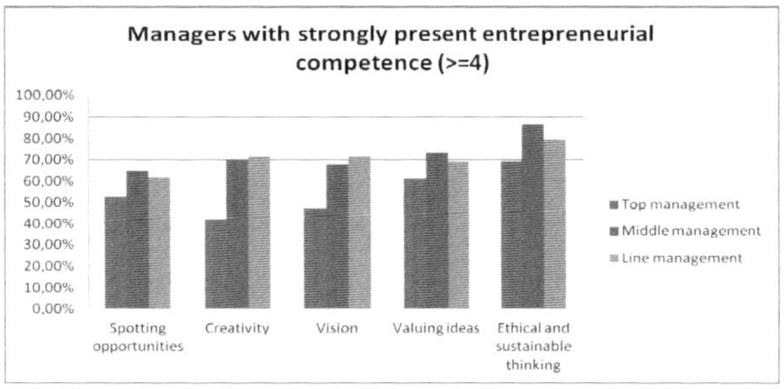

Figure 3. Percentual presence of managers with strongly present entrepreneurial competence, by managerial level

Out of the three analyzed categories of managers, in all five entrepreneurial competences, the smallest presence of "strongly present competence" is evidenced in the category of the top managers. In this category of managers, competence *spotting opportunities* is strongly present at 52.8% of the participants, competence *creativity* at 41.7%, *vision* at 47.2%, *valuing ideas* at 61.1% and *ethical and sustainable thinking* at 69.4%. On the other hand, the results of middle and line managers are higher and comparable.

Among middle managers, competence *spotting opportunities* is strongly present at 64.9% of the participants, competence *creativity* at 70.3%, *vision* at 67.6%, *valuing ideas* at 73% and *ethical and sustainable thinking* at 86.5%. Finally, in the category of the line managers, competence *spotting opportunities* is strongly present at 61.5% of the participants, competence *creativity* at 71.8%, *vision* at 71.8%, *valuing ideas* at 69,2% and *ethical and sustainable thinking* at 79.5%.

Based on the EntreComp model (Bacigalupo et al., 2016), the competence *spotting opportunities* includes capacity to create, identify and seize the opportunities for values creation by exploring the economic landscape, with identification of needs and challenges that should be met while analysing the context and establishing all necessary connections to create opportunities for value creation. Further, *creativity* as an entrepreneurial competence understands capacity for development of different ideas and opportunities for value creation including better solution for existing and future challenges, practicing new, innovative approaches and combination of knowledge and approaches for achieving valuable effects. Finally, *vision* as a competence integrates the capacity for imagination and visualisation of different future scenarios to turn the ideas into action, aiming value creation. Having in mind the above presented results it can be concluded that among the companies, participants in this research, the creative power for new ideas and opportunities is concentrated among the middle and line managers. This can be supported by the fact that the line and middle management have close contact on a daily bases with the customers on one side, thus can directly listen and assess the customers' needs. On the other hand, the lower managerial levels are in more direct contact with employees as well; therefore they can mutually exchange and discuss the new ideas and opportunities. Taking into account the existing presence of these competences among line and middle management it can be concluded that there is a fertile soil, the ideas generated by the employees to be supported and promoted to the higher managerial levels.

According to the EntreComp conceptual model, the competence *valuing ideas* consists of capacity to judge and recognise the potential and value of the ideas and to share and protect them. The quantity of the managers with strong presence of this competence is comparable among the different levels of the managers, although the most dominant at middle management level. These results support the fact that once the new ideas are generated or new opportunities spotted by the lower hierarchical levels in the companies there is a capacity by the higher managerial levels the ideas to be correctly valued and upon the best ones correspondent actions to be taken.

Among all three managerial levels there is significant number of managers with strong presence of the competence *ethical and sustainable thinking*, with largest presence among the middle managers and smallest among top managers. However, it can be concluded that the managers of all levels behave ethically, think sustainably, access the impact and are accountable for all taken steps not only for generation of the new ideas and opportunities, but for the impact of the taken action for implementation of those ideas as well.

Derived on the above presented findings, it can be summarized that among the managers, participants in the research, there is a high level of presence of the entrepreneurial competences belonging to the category "Ideas and opportunities" of the EntreComp model. This means that the managers would be the main driving force, facilitator and immense support of the process of new opportunities identification, appraisal and seize as well as on creativity and innovation. Additionally, support of these activities in the same time means support of the corporate entrepreneurial orientation. This is in line with the researches of other scholars who stress the importance of managers and their support as crucial factors for corporate entrepreneurship, particularly in the area of the new ideas and opportunities. In that direction Heavey and Simsek (2013) recognize that the main burden of the corporate entrepreneurship is on the top management level, while Hunter et al. (2007) identify top-management support, as well as support for discussion about new ideas and availability of resources for their implementation, as key factors for corporate entrepreneurship. Many authors recognize the middle managers' crucial role in support and promotion of new ideas and opportunities. Kanter (1985) endorses the middle managers for creating the positive climate for new, innovative ideas and decreasing resistance to change thus enabling the corporate entrepreneurship to develop, while Burgelman (1984) recognize the importance of the middle managers in communication of the new ideas to the top managers. Finally, Kuratko and Audretsch (2013) debate that "the middle management level is where entrepreneurial opportunities are given the best chance to flourish based on the resources likely to be deployed in their pursuit" (p.327).

6 Conclusion

The presence of five entrepreneurial competences belonging to the "Ideas and opportunities" area of the EntreComp conceptual model among the managers in selected companies in the Republic of Macedonia was explored. The results show relatively high presence of all competences among the managers, leading to a conclusion that there is a capacity among the managers to support the entrepreneurial behaviour in their companies in the area of new ideas and opportunities. Hence, they are creating conditions for development of the corporate entrepreneurship. The analysis of the number of managers with strongly present entrepreneurial competences by managerial level led to a conclusion that the creative power for innovations, new ideas and opportunities is concentrated at the lower managerial levels, therefore the new ideas can be expect to follow bottom-up approach. It means, the ideas will be created at the middle and line management levels, but assessed and approved on all managerial levels, ending at the top managerial level. Finally, ethical and sustainable thinking regarding the consequences and the impacts of the new ideas was highly present among the managers of all managerial levels.

References

Ahmad N. and Hoffman, A. (2007). *A Framework for Addressing and Measuring Entrepreneurship*. Entrepreneurship Indicators Steering Group, OECD, Paris.

Amabile, T. M., Schatzel, E. A., Moneta, G. B. and Kramer, S. J. (2004).Leader behaviours and the work environment for creativity: Perceived leader support. *Leadership Quarterly*, 15, pp.5-32.

Antoncic, B. and Hisrich, R. D. (2001). Intrapreneurship: Construct refinement and cross-cultural validation. *Journal of Business Venturing*, 16(5), pp.495-527.

Antoncic, B. and Hisrich, R. D. (2003). Clarifying the intrapreneurship concept. *Journal of Small Business and Enterprise Development*, 10(1), pp.7-24.

Bacigalupo, M., Kampylis, P., Punie, Y., Van den Brande, G. (2016). *EntreComp: The Entrepreneur-ship Competence Framework*. Luxembourg: Publication Office of the European Union; EUR 27939 EN; doi:10.2791/593884

Baron, R. A. (1998). Cognitive mechanisms in entrepreneurship: Why and when entrepreneurs think differently than other people. *Journal of Business Venturing*, 13(4), pp.275-294.

Buchanan, J. M. and Vanberg, V. J. (1991). The Market as a Creative Process. *Economics and Philosophy*, 7(2), pp.167-186.

Burgelman, R. A. 1984. Designs for Corporate Entrepreneurship In Established Firms, *California Management Review*, 26 (3), pp.154-166.

Bygrave, W. D. and Hofer, C. W. (1991). Theorizing about entrepreneurship. *Entrepreneurship: Theory & Practice*, 16(2), pp.13-22.

Drucker, P F. (1985). *Innovation and Entrepreneurship: Practice and Principles*. New York, USA: HarperBusiness.

Gartner, W. B., Shaver, K. G., Gatewood, E., and Katz, J. A. (1994). Finding the entrepreneur in entrepreneurship (editorial). *Entrepreneurship: Theory & Practice*, 18(3), pp.5-9.

Gartner, W. B. (1988). Who is an entrepreneur? is the wrong question. *American Journal of Small Business*, 12(4), pp.11-32.

Heavey, C., and Simsek, Z. (2013). Top management compositional effects on corporate entrepreneurship: The moderating role of perceived technological uncertainty. *Journal of Product Innovation Management*, 30(5), pp.837–855.

Hunter, S.T., Bedell, K.E., and Mumford, M.D. (2007). Climate for Creativity: A Quantitative Review. *Creativity Research Journal*, 19(1), pp.69–90.

Ireland, R. D., Covin, J. G., Kuratko, D. F. (2009).Conceptualizing corporate entrepreneurship strategy. *Entrepreneurship Theory and Practice*, 33, pp.19-46.

Kanter, R. M. 1985. Supporting Innovation and Venture Development in Established Companies, *Journal of Business Venturing* ,1(1), pp. 47-60.

Kirzner, I. M. (1978). *Competition and entrepreneurship*. Chicago: University of Chicago Press.

Kuratko, D. F., Hornsby, J. S. and Bishop, J. W. (2005). An examination of managers' entrepreneurial actions and job satisfaction. *The International Entrepreneurship and Management Journal*, 1(3), pp.275-291.

Kuratko, D.F. and Audretsch, D.B. (2013). Clarifying the domains of corporate entrepreneurship, *International entrepreneurship and management journal*, 9(3), pp. 323-335.

Low, M.B., and MacMillan, I.C. (1988). Entrepreneurship: Past research and future challenges. *Journal of Management* 14(2),pp.139-161.

Lukeš, M., Stephan, U. and Černíková, A. (2009). *Measuring innovative behavior and innovation support*. Paper presented at 2nd ISPIM Symposium, New York.

Lukeš, M. (2012) Supporting Entrepreneurial Behaviour and Innovation *Organizations Central European Business Review*, 2012(2), pp.29-36.

Lumpkin, G. T. (2007). Intrapreneurship and innovation. In Baum, J. R., Frese, M., Baron, R. (Eds.), *The Psychology of Entrepreneurship*, pp.237-264. Mahwah: Lawrence Erlbaum Associates.

Morris, M. H., Kuratko, D. F. (2002). *Corporate entrepreneurship - entrepreneurial development within organizations*. Fort Worth, Texas: Harcourt, Inc.

Nielsen, R.P., Peters, M.P. and Hisrich, R.D. (1985), Intrapreneurship Strategy for Internal Markets-Corporate, Non profit and Government Institution Cases. *Strategic Management Journal*. 6(2), pp181-189.

Pinchot, G. (1985). *Intrapreneuring: Why You Don't Have to Leave the Organization to Become an Entrepreneur*. New York: Harper & Row.

Pinchot, G. (2010), *The Pinchot Perspective. Back to Intrapreneuring* [Online], Available: http://www.pinchot.com/2010/01/back-to-intrapreneuring.html, accessed 09.02.2017

Robinson, A. G. and Schroeder, D. M. (2006). *Ideas are free: how the idea revolution is liberating people and transforming organizations*. San Francisco: Berrett- Koehler Publishers.

Sathe, V. (2003). *Corporate entrepreneurship*. Cambridge University Press.

Stevenson, H. H. and Jarillo, J. C. (1990). A paradigm of entrepreneurship: Entrepreneurial management. *Strategic Management Journal*, 11(5), pp.17-27.

Saunders, M., Lewis, P. and Thornhill, A. (2003). *Research Methods for Business Students,* 3th edition. Harlow, Pearson Education Limited.

Schumpeter, J., (1934). *The Theory of Economic Development: An Inquiry into Profits, Capital, Credit, Interest and the Business Cycle*. Cambridge, MA: Harvard University Press.

Scott, S. G. and Bruce, R. A. (1994). Determinants of innovative behaviour: A path model of individual innovation in the workplace. *Academy of Management Journal*, 37 (3), pp.580-607.

Shane, S. A. (2000). Prior knowledge and the discovery of entrepreneurial opportunities. *Organization Science,* 11(4), pp.448–469.

Zahra, S.A. (1993). A conceptual model of entrepreneurship as firm behaviour: A critique and extension. *Entrepreneurship Theory and Practice* 17(4), pp.5-21.

AUTHOR INDEX

Majan University College
Mutrah | Way No. 2621 | Building No. 1986
P. O. Box: 710, P.C.: 112
Ruwi | Sultanate of Oman

ISBN 978-1-5108-6299-9

9 781510 862999

CURRAN ASSOCIATES INC.
proceedings.com